Thyroid Gland Development and Function

....

Endocrine Development

Vol. 10

Series Editor

Martin O. Savage *London*

Thyroid Gland Development and Function

Volume Editors

Guy Van Vliet *Montreal, Que.*
Michel Polak *Paris*

41 figures, 10 in color, and 11 tables, 2007

KARGER

Basel · Freiburg · Paris · London · New York ·
Bangalore · Bangkok · Singapore · Tokyo · Sydney

Guy Van Vliet, MD
Endocrinology Service and Research Center
Department of Pediatrics
Sainte-Justine Hospital
University of Montreal
Montreal, Quebec, Canada

Michel Polak, MD, PhD
Endocrinologie pédiatrique
INSERM U845
Hôpital Necker Enfants Malades
Paris, France

Library of Congress Cataloging-in-Publication Data

Thyroid gland development and function / volume editors, Guy Van Vliet,
Michel Polak.
p. ; cm. – (Endocrine development, ISSN 1421-7082 ; v. 10)
Includes bibliographical references and indexes.
ISBN-13: 978-3-8055-8296-4 (hard cover : alk. paper)
1. Thyroid gland–Diseases. 2. Thyroid gland–Pathophysiology.
3. Thyroid gland–Growth. I. Van Vliet, Guy. II. Polak, Michel. III. Series.
[DNLM: 1. Thyroid Gland–growth & development. 2. Thyroid
Gland–physiology. 3. Thyroid Gland–physiopathology. 4. Thyroid
Diseases–genetics. W1 EN3635 v.10 2007 / WK 200 T5472 2007]
RC655.T4844 2007
616.4′4–dc22

2007019947

Bibliographic Indices. This publication is listed in bibliographic services, including Current Contents® and Index Medicus.

www.karger.com
Printed in Switzerland on acid-free and non-aging paper (ISO 9706) by Reinhardt Druck, Basel
ISSN 1421–7082
ISBN 978–3–8055–8296–4

Contents

Foreword

This volume in the *Endocrine Development* series entitled *Thyroid Gland Development and Function* fits perfectly into the primary aim of the series, which is to discuss the physiology and clinically relevant pathophysiology of key endocrine systems. Scientific and clinical interests are given prominence in this volume. Professor Polak and Professor Van Vliet are highly experienced, both from experimental and clinical standpoints, to edit this issue. They have chosen subjects of major interest and contributors of very high quality.

Human thyroid development and its defects are described with the help of genetic studies in mouse models. The metabolic aspects of thyroid hormone action are also discussed. Genetic defects of thyroid hormone synthesis are covered and their clinical relevance debated. The important field of thyroid cancer in the context of spontaneous occurrence and as part of familial neoplasia syndromes is described in detail. Finally the important problem of environmental iodine deficiency which has emerged as a global public health concern is rightly included.

Overall, this excellent volume will inform scientists and clinicians of key areas in the field of thyroid disorders. I enthusiastically welcome this latest addition to the series.

Martin O. Savage, London

Preface

In 1985, Karger published a book entitled *Pediatric Thyroidology* which had been edited by F. Delange, D.A. Fisher and P. Malvaux. Since then, tremendous advances have taken place in developmental and molecular biology. These advances have had a major impact on all fields of medicine, and pediatric thyroidology is no exception. Consistent with its publication in the *Endocrine Development* series edited by Martin Savage, this book starts with chapters focusing on developmental abnormalities of the thyroid gland in genetically engineered mice. Studies are described on the possible mendelian and non-mendelian mechanisms involved in abnormalities of thyroid development in humans and on their proper classification by imaging. Ironically, the commonest developmental abnormality, a defect in migration resulting in thyroid ectopy, remains an enigma in the field.

The recent advances in our understanding of thyroid hormone metabolism and transport into the cells, which have been revealed by astute observations of 'experiments of nature' observed in children, followed by sophisticated molecular investigations, are reviewed next.

What makes the thyroid rather unique in the field of endocrinology is its critical dependence on an environmental factor, iodine. Pregnant women are particularly sensitive to a low nutritional supply of iodine. Within the thyroid, iodine needs to be oxidized, a process which requires H_2O_2; genetic lesions resulting in decreased function of the protein involved in the generation of H_2O_2 lead to a form of hypothyroidism that may be exacerbated during pregnancy and the newborn period. The intricate relationships between maternal and fetal thyroid function may result in major consequences of maternal hypothyroidism on the psychomotor development of the offspring. These aspects are reviewed in the next section.

The biology of tumors arising from thyroid follicular cells in childhood differs from those arising from the same cells later in life. Tumors arising from the parafollicular or C cells represent the first example of the major impact that DNA-based diagnosis has had on the practice of pediatric endocrinology; in this area, we stand on the verge of 'codon-specific' medicine. These pediatric thyroid tumors are reviewed in the last section of this book.

We are well aware that our choice of topics may seem rather arbitrary. It was not our aim to produce a complete overview of pediatric thyroid diseases and their consequences, but rather to focus on selected topics which fell under the general umbrella of *Endocrine Development* and in which we felt that major advances had recently been made, usually through a combination of clinical observations and patient-oriented basic biological investigations. We thank all the authors for their outstanding contributions and sincerely hope the readers will learn from perusing this book as much as we have from editing it.

Guy Van Vliet, Montreal, Que.
Michel Polak, Paris

Van Vliet G, Polak M (eds): Thyroid Gland Development and Function.
Endocr Dev. Basel, Karger, 2007, vol 10, pp 1–14

Murine Models for the Study of Thyroid Gland Development

Mario De Felice[a,b], *Roberto Di Lauro*[a,b]

[a]Dipartimento di Biologia e Patologia Molecolare e Cellulare,
Università Federico II, Napoli, e [b]IRGS, Biogem s.c.a r.l., Ariano Irpino (AV), Italia

Abstract

Gene targeting technology has allowed the generation of mouse mutants which lack specific genes. These mice represent a valuable tool for identifying the in vivo functions of proteins and for dissecting the pathways that control the development and differentiation of the numerous structures of the body. What we know about thyroid morphogenesis is largely due to studies on murine models generated in gene-targeting experiments. Although several points remain to be elucidated, a number of genes involved in thyroid organogenesis have been identified in recent years. In addition, this information has greatly improved our knowledge of the pathogenetic mechanisms of thyroid dysgenesis in humans. This review summarizes the complex processes leading to thyroid development mostly by describing the phenotype of currently available knockout animals.

Introduction

Fifteen years ago it was reported that transcription factors, identified for their relevance in the expression of genes specific for differentiated thyrocytes, were present in the thyroid primordium [1]. This discovery made it possible to begin the exploration of the genetic basis of thyroid development. In recent years genes and mechanisms involved in thyroid morphogenesis have been partially identified. Many of the steps of thyroid development have been elucidated, but a number of crucial issues of this process are still obscure. At the same time, the generation of murine strains in which thyroid-relevant genes have been disrupted has provided useful animal models of human thyroid dysgenesis (TD), the most frequent cause of congenital hypothyroidism [2]. These models have been valuable in elucidating the molecular pathology of TD, confirming that it is a genetically heterogeneous disease. In addition, the study of patients affected by TD is providing insights into the molecular mechanisms involved in normal thyroid development.

The aim of this review is to summarize what we know to date and to highlight what is still unknown about the processes regulating the normal and disturbed development of the thyroid, mostly as deduced from the phenotype of knockout animals.

Genetics of Early Stages of Thyroid Development

Specification of the Thyroid Anlage

In mammals the developing thyroid is first visible as a thickening of the endodermal epithelium emerging at the most anterior part of the foregut. This structure, called thyroid anlage, is evident by E8–8.5 in mice and by E20–22 in humans. The endodermal cells of the thyroid anlage, acquiring a specific molecular signature – the co-expression of the four transcription factors *Hhex* [3], *Titf1* [1], *Pax8* [1] and *Foxe1* [4] – distinguish themselves from their neighbors. The process leading to the establishment of the thyroid anlage is defined as thyroid specification; it is one of the effects of the morphogenetic events that regionalize the undifferentiated original endodermal tube into anatomically defined compartments which undertake a defined developmental program [5]. The 'prime mover' of the initial specification of the thyroid anlage is obscure. It has been hypothesized that the cells are destined to a thyroid fate as a consequence of short-range inductive signals originating either from the surrounding mesenchyme or from the endothelial lining of the adjacent aortic sac. However, the molecules involved in this process are unknown.

Defects in thyroid specification should result in thyroid agenesis, i.e. total absence of the gland as a consequence of impaired organogenesis. Any gene that regulates formation of the foregut, such as Nodal, transcription factors downstream of Nodal signaling, FGF4, members of the GATA family or Sox genes, could play a role in the initial specification of the thyroid. However, mutant mice with targeted inactivation of genes involved in this process are not informative in dissecting a thyroid-specific network. Actually, in most cases, these mice show developmental arrest at stages that preclude assessment of thyroid specification. Other candidate genes may be those encoding factors responsible for the onset of *Titf1*, *Pax8* and *Hhex* expression in the thyroid bud. These genes are still to be identified and how they affect thyroid specification is still to be determined [6]. Up to now, we do not have any murine model that displays bona fide thyroid agenesis due a defective initiation of thyroid morphogenesis.

Table 1. Different phases of thyroid development in mice: morphological features, expression of relevant genes and capacity to produce thyroid hormones

Embryonic day	Stages of morphogenesis	Controller genes		Functional differentiation		Thyroid hormones
		Titf1 *Foxe1* *Pax8* *Hhex*	*Ffgr2*	*Tg* *TPO* *Tshr*	*NIS*	
E8	undifferentiated endoderm	–	–	–	–	–
E8.5	thyroid anlage appears	+	–	–	–	–
E9.5	thyroid bud evaginates	+	–	–	–	–
E10.5	thyroid bud migration begins	+	–	–	–	–
E11.5	thyroglossal duct disappears	+	+	–	–	–
E12.5	expansion of thyroid primordium	+	+	–	–	–
E13.5	thyroid migration is complete	+	+	–	–	–
E14.5–15	definitive bilobed shape	+	+	+	–	–
E15.5–16	onset of folliculogenesis	+	+	+	+	+
E16.5	terminal differentiation	+	+	+	+	+
E17–18.5	expansion of the thyroid	+	+	+	+	+

Budding of the Thyroid Primordium

By E9.5, the thyroid anlage evaginates from the floor of the pharynx, invades the surrounding mesenchyme and forms a definite bud that appears as a flask-like structure at E10.5 (table 1). It has been recently demonstrated that cell proliferation is low or lacking in the thyroid bud [7]. These data have been used to suggest that to some extent the growth of the thyroid primordium could be due, at least in part, to the recruitment of cells from the pharyngeal endoderm. The bud assumes an elongated shape and it rapidly becomes an endodermal-lined diverticulum that begins to migrate caudally into the mesenchyme. The thyroid primordium is still connected to the epithelium of the pharynx by a narrow structure, the thyroglossal tract, that gradually regresses and by E11.5 disappears. Thyroid morphogenesis follows the same pattern in humans [8], in whom the thyroid diverticulum begins its migration at E26 and loses its continuity with the pharynx at E37.

Disturbances in the genetic network controlling any step following the specification of the thyroid anlage could impair survival or proliferation of the thyroid precursor cells causing athyreosis, a dysgenesis characterized by the disappearance of the thyroid primordium and, as consequence, of the adult gland. Thyroid precursor cells are hallmarked by the presence of the transcription factors Hhex,

Titf1, Pax8 and Foxe1; each of them plays an essential role in the morphogenesis of the gland [6]. Accordingly, mice carrying null mutations in each of these genes are good models of athyreosis. Actually, in these mice, the morphogenesis of the gland begins but the thyroid bud disappears. *Hhex*, *Titf1* and *Pax8* mutants will be discussed below, while the features of *Foxe1* null mice will be described in the next section.

Hhex

Hhex (formerly known as Hex for hematopoietically expressed homeobox or Prh for proline-rich homeobox) is a homeodomain-containing transcription factor, first identified in multipotent hematopoietic cells. It is encoded by a gene located on chromosome 19 in mice and chromosome 10q23.32 in humans. In embryos, at early stages of development, Hhex is detected in the primitive and definitive endoderm. It is then expressed in the primordium of several organs derived from the foregut, including the thyroid bud [3].

Studies of *Hhex*$^{-/-}$ embryos [9] show that this protein plays a critical role in the development of the liver, forebrain, heart and thyroid. In *Hhex* null embryos at E9, the thyroid anlage is present and the expression of *Titf1*, *Pax8* and *Foxe1* is not affected. At E10, in the absence of Hhex, thyroid budding is severely impaired and the thyroid primordium is represented only by a few nonmigrating cells which do not express *Titf1*, *Pax8* or *Foxe1* mRNA. At later stages, the primordium disappears. These data strongly suggest that Hhex has no role in thyroid specification but is involved in the survival of already determined thyroid precursors. Since Hhex is required to maintain *Titf1*, *Pax8* and *Foxe1* expression in the developing thyroid [6], we cannot exclude that the absence of these factors is the direct cause of the thyroid phenotype displayed by *Hhex*$^{-/-}$ embryos.

No *HHEX* mutations in humans have been described so far.

Titf1

Titf1 (formerly known as TTF-1 for thyroid transcription factor-1, or Nkx2-1 or T/EBP) is a homeodomain-containing transcription factor member of the Nkx2 family, initially identified as a protein able to bind to specific sequences in the thyroglobulin (*Tg*) and thyroid peroxidase (*TPO*) promoters. Titf1 is encoded by a gene located on chromosome 12 in mice and on chromosome 14q13 in humans. During embryonic life, in addition to the thyroid primordium, Titf1 is detected in the endodermal cells of trachea and lungs and in selected areas of the forebrain, including the developing posterior pituitary [1]. In the developing thyroid, *Titf1* is expressed in the precursors of both follicular and C cells and in the epithelial cells of the ultimobranchial body [10].

Mice carrying a null mutation for the *Titf1* gene die at birth and display hypoplastic lungs, alterations in the forebrain, lack of pituitary and thyroid [11].

The thyroid anlage is comparable in wild-type and *Titf1*$^{-/-}$ embryos up to E9. However, already at E10.5 the thyroid primordium shows a reduced expression of *Pax8*, *Foxe1* and *Hhex* [6] and appears much smaller in size compared to wild type. Subsequently, thyroid cells disappear probably through apoptosis. Consistently with the expression pattern of Titf1, in the absence of this factor, calcitonin-producing C cells and epithelial cells of the ultimobranchial body disappear. The ultimobranchial body is correctly formed but undergoes apoptotic degeneration by E12 in the early phase of migration [12]. It is worth noting that the thyroid parenchyma is composed of three epithelial cell populations of different embryological origin – follicular, parafollicular and ultimobranchial body-derived cells; Titf1 is dispensable for the initial specification but is absolutely required for the survival of all three cell types. Titf1 functions are in part dosage-sensitive. Indeed *Titf1*$^{+/-}$ mice display decreased coordination, mild hyperthyrotropinemia [13] and an abnormal fusion of the ultimobranchial body with the thyroid diverticulum [12].

The genetic pathway controlled by Titf1 in the thyroid primordium is still elusive, even if we know that the absence of this transcription factor abolishes the expression of *Bmp4* and *Fgf8* in the developing lungs and in the posterior pituitary, respectively. Titf1 could regulate the survival of the thyroid precursor cells through Fgf-dependent mechanisms. Consistent with this hypothesis are the findings that *Fgfr2* is expressed in the thyroid bud starting at E11 and that mice deficient for this receptor lack a thyroid gland [14].

In humans, *TITF1* loss-of-function mutations present a remarkable dominant effect. These patients are affected by a syndrome characterized by neurological disturbances (choreoathetosis) [13, 15], respiratory distress and generally mild hypothyroidism, while scintigraphy shows variable results, ranging from normal to absent uptake.

Pax8

Pax8 (paired box gene 8) is a member of a family of transcription factors characterized by the presence of a DNA binding domain called paired domain encoded by a gene located on chromosome 2 in both humans and mice. Pax8 recognizes and binds to specific sequences present in both *Tg* and *TPO* promoters and, in differentiated thyroid follicular cells, directly interacts with Titf1. This cooperation could be relevant in the stimulation of thyroid genes. During embryonic life, *Pax8* is expressed in the myelencephalon, in the kidneys and in the endodermal cells of the developing thyroid since E8.5 [16].

Pax8$^{-/-}$ embryos show a thyroid anlage which cannot be distinguished morphologically from that of the wild type. The thyroid bud evaginates from the endoderm and migrates into the mesenchyme. However, in absence of Pax8 by E11.5 the thyroid primordium appears hypoplastic [10] and does not express

Foxe1 and *Hhex* [6]. A day later, the follicular cells are essentially undetectable. *Pax8*$^{-/-}$ pups present a rudimentary gland, composed almost completely of calcitonin-producing C cells and die within 2–3 weeks of birth. Thus Pax8, like Titf1, is required for the survival of thyroid cell precursors and to maintain the expression of other thyroid-specific regulatory genes.

In humans, individuals carrying heterozygous loss-of-function mutations in the *PAX8* gene show hypothyroidism with TD [17]. Thyroid alterations are variable, from mild hypoplasia of the gland to absence of the thyroid.

Migration of the Thyroid Primordium

By E12 proliferative activity is detected in the thyroid primordium which begins to expand laterally; at E13–14 it reaches its definitive pretracheal position where it merges with the ultimobranchial bodies containing the precursors of C cells derived from the neural crest. In humans, the thyroid primordium reaches its destination, anterior to the trachea and inferior to the cricoid cartilage, by E44–48, after a 'journey' requiring almost 4 weeks.

The genetic basis of the migration of the thyroid primordium has been only partially elucidated. It is hard to attribute a role to Hhex, Titf1 or Pax8: on one hand, in *Hhex* or *Titf1* null embryos thyroid morphogenesis stops before the start of the migration; on the other hand, in the absence of Pax8, the thyroid primordium correctly migrates into the mesenchyme. In contrast, Foxe1, although expressed along the entire endodermal epithelium of foregut, has a specific function in this process.

Foxe1 (formerly called TTF-2 for thyroid transcription factor-2) is a transcription factor, a member of the winged helix/forkhead family, encoded by a gene located on chromosome 4 in mice and on chromosome 9q22 in humans. At an early stage of embryonic life, it is expressed in the developing thyroid, Rathke's pouch, tongue and esophagus; at a later stage, it is detected in the secondary palate, definitive choanae, whiskers and hair follicles [18]. Analysis of *Foxe1* null mice shows that in the absence of this factor the specification of the thyroid anlage is correct. However, the migration of thyroid precursor cells is impaired: at E10 in *Foxe1*$^{-/-}$, the thyroid primordium is still on the floor of the pharynx while in wild-type embryos it is already descending towards its final location. At later stages, *Foxe1* null thyroid cells either disappear or form an ectopic small thyroid remnant able to synthesize Tg [19].

These data indicate that Foxe1, in addition to cooperating in the control of the survival of thyroid cells, is specifically involved in the migration of the thyrocytes. Its crucial role in promoting thyroid migration is confirmed by studies on another mouse model where the expression of this factor is restricted only to the developing thyroid. In such a mutant mouse, the thyroid bud migrates, demonstrating that this phenomenon is a cell-autonomous event that depends on Foxe1-controlled features intrinsic to the thyroid precursor cells [6]. However, genes that accomplish the migration program remain to be identified. Furthermore, in addition to these mechanisms, other morphogenetic events occurring in the neck region and in the mouth can contribute to drive the thyroid primordium towards its final location [20].

Homozygous defects in the *FOXE1* gene in humans are associated with Bamforth syndrome, characterized by cleft palate, bilateral choanal atresia, spiky hair and athyreosis [21].

Gene Interactions at Early Stages of Thyroid Morphogenesis

The phenotype of the knockout mice summarized above not only demonstrates that Titf1, Hhex, Pax8, and Foxe1 are required for correct thyroid development but also indicates that these genes are linked in a complex network of reciprocal regulatory interactions. In the thyroid anlage the expression of *Titf1*, *Hhex*, and *Pax8* is not dependent on the expression of each other, since at E9 in each knockout mouse the other two factors are unaffected. In contrast, at a later stage, each of them controls the maintenance of the expression of the others [6]. For this reason, we cannot exclude that the athyreosis displayed by each of these mutants is due to the removal of the entire regulatory network. Interestingly, Foxe1 holds a lower position in the genetic regulatory cascade controlling thyroid development. Actually, the simultaneous presence of Titf1, Hhex and Pax8 is required for its expression, while in the *Foxe1* null mouse, *Titf1*, *Hhex* and *Pax8* are correctly expressed in the thyroid bud. These data are consistent with the finding that in the developing human thyroid the expression of both *TITF1* and *PAX8* precedes the onset of *FOXE1* expression [8].

Genetics of Late Stages of Thyroid Development

Lobulation

At E14–15 the developing thyroid is composed of a semicircular midline portion and two rudimentary paratracheal lobes. Then, the lateral lobes enlarge; as a consequence of this process the gland assumes its definitive shape: two lobes connected by a narrow isthmus. At the same time, the thyroid parenchyma begins to reorganize into cords of cells and small rudimentary follicles become evident; in addition migrating C cells, derived from the ultimobranchial bodies, disseminate into the parenchyma and eventually assume a parafollicular position. In humans, around E50 the thyroid is already separated into two lobes and begins to form rudimentary follicles.

When the process of lobulation is disturbed, the correct organogenesis of the gland is impaired: the thyroid fails to separate into two symmetric lobes and forms a unique mass (hemiagenesis). Some animal models are revealing themselves to be a useful tool to dissect the mechanisms controlling the formation of the lobes.

Sonic hedgehog (Shh) is a member of the hedgehog family, soluble ligands for Patched receptor, encoded by a gene located on chromosome 5 in mice and on chromosome 7q36 in humans. In embryos, *Shh* is expressed in many tissues derived from all three germ layers. Remarkably, it is detected along the entire foregut epithelium including the pharyngeal floor but is excluded from cells of the thyroid anlage [6]. The complex phenotype of $Shh^{-/-}$ embryos testifies that this molecule is a key regulator of embryogenesis. Recent studies on these mutants have proved a role of Shh in thyroid organogenesis. In *Shh* null embryos the early steps of thyroid morphogenesis are unaffected but the whole lobulation process seems to be impaired: at E15, the developing thyroid appears as a single midline tissue mass which is located laterally to the trachea at the end of organogenesis [22]. Since neither Shh nor its receptor have been detected in thyroid cells, these data strongly suggest that the lobulation process is instructed by other structures whose correct patterning is disturbed in the absence of Shh. Candidate structures could be vessels located close to the thyroid tissue which display an aberrant development in *Shh* null embryos. This hypothesis is consistent with the report of thyroid hemiagenesis in patients affected by diseases characterized by congenital anomalies of the heart and great vessels, such as the Di George or truncus arteriosus syndrome. However, the genetic basis of the formation of symmetric lobes is far from being entirely elucidated. Actually, a high frequency of thyroid hemiagenesis has also been reported in mice double heterozygous for the null allele of *Titf1* and *Pax8* [23], genes which are not expressed in structures close to the developing thyroid. Thus autonomous events restricted to the thyroid cells must interact with signals from adjacent tissues to complete the lobulation process.

Functional Differentiation

By E14.5 thyroid cells go through a differentiative program that is completed, 2 days later, with the synthesis of thyroxine. The functional differentiation of thyroid cells is a consequence of the expression, according to a precise temporal pattern, of a series of proteins essential for thyroid hormone biosynthesis: Tg, TPO, TSH receptor (Tshr), sodium/iodide symporter (NIS), thyroid oxidases (Thoxs) and pendrin (PDS). The mechanisms controlling the functional differentiation of thyroid cells at this stage are under investigation. In adults, the most important regulator of thyroid physiology is TSH, acting through its receptor Tshr; how and to what extent TSH/Tshr signaling is involved in thyroid cell differentiation has been only recently studied in animal models.

Tshr (thyroid-stimulating hormone receptor) is a member of the family of G protein-coupled receptors encoded by a gene localized on chromosome12 in mice and 14q31 in humans. Tshr is detected in thyroid cells by E15 – at the same time as when TSH is produced by the fetal pituitary – and its expression strongly increases by E17. The availability of mice that carry mutations in the *Tshr* gene has provided a powerful tool to elucidate the role of TSH/Tshr signaling during embryonic life. Both *Tshr*^{hyt}/*Tshr*^{hyt} (carrying a spontaneous loss-of-function mutation in the *Tshr* gene) [24] and *Tshr*$^{-/-}$ mice [25] (in which the *Tshr* gene has been disrupted by gene targeting) display severe hypothyroidism after birth. A detailed analysis on E17 embryos has shown that in the absence of a functional Tshr, NIS and TPO are almost undetectable while the expression of Tg appears to be only slightly decreased in the mutant thyroid in comparison with wild-type ones. These data indicate that the TSH/Tshr pathway plays an essential role in the differentiative program of the thyroid cell exerting a coordinated and tight control of two proteins that are key to the process of Tg iodination.

Among the molecular defects causing TD in humans, mutations in the *TSHR* gene represent the most frequent finding. Individuals homozygous or compound heterozygous for loss-of-function mutations in the TSHR gene show mild or severe thyroid hypoplasia [26]. In many cases, the thyroid size appears normal and only increased TSH levels characterize these subjects.

Expansion of the Fetal Thyroid

In the last stages of embryonic life thyroid cells show a high proliferation rate so that the thyroid increases in size. At the same time, the gland develops its peculiar highly organized architecture; by E17.5, the gland is composed of small follicles accumulating Tg in the lumen and surrounded by a capillary network. In the mouse, the regulation of growth and function of the thyroid by the hypothalamic-pituitary axis is fully active only after birth. In humans, the thyroid displays an evident follicular organization after 10–11 weeks of gestation but the gland continues to grow until term; furthermore, the hypothalamic-pituitary-thyroid axis is operative at midgestation.

In adult mice, the TSH-induced cAMP pathway is the main regulator of thyroid growth. This is confirmed by the hypoplastic adult thyroid displayed by all animal models carrying natural or induced mutations in *Tshr* or its cognate ligand. In addition, transgenic mice overexpressing in the thyroid the A_2 adenosine receptor, which causes a constitutive activation of adenylyl cyclase, show a dramatic hyperplasia of the gland [27]. In contrast, at E17, in the absence of a functional Tshr, the size of the gland and number of proliferating thyreocytes are comparable in mutant and wild-type embryos [24]. Furthermore, at birth, the thyroid in A_2 adenosine receptor transgenic mice is comparable to that of wild-type newborn mice [27]. These data indicate that during embryonic life the growth of the gland is controlled by mechanisms independent of TSH/Tshr/cAMP signals. What these mechanisms could be is still puzzling. Interestingly, mice double heterozygous for the null allele of *Titf1* and *Pax8* show an impaired organogenesis of the thyroid: in many cases a hypoplastic thyroid is evident at E15 and hypoplasia persists at birth. It is possible that growth factors controlled by both Titf1 and Pax8 are involved in the proliferation of immature thyroid cells.

From Mouse Models to Human Diseases

TD is the result of disturbances in the migration, growth and/or differentiation of the thyroid primordium. The term includes different entities: a gland located in an abnormal position (ectopia), complete absence of the thyroid (athyreosis), and a thyroid of decreased size (hypoplasia).

The discovery that the disruption of genes involved in thyroid morphogenesis causes an impaired development of the gland in mice has led researchers to look for mutations in the homologous human genes in patients affected by congenital hypothyroidism with TD. This effort has been successful and several groups report a number of cases of congenital hypothyroidism associated with mutations in either *TITF1*, *PAX8*, *FOXE1* or *TSHR*, proving that TD can be a genetic and inheritable disease. The phenotype of human diseases is generally close to that of the corresponding murine models. However, some differences should be emphasized.

(a) Mutations thus far identified account only for a very small number of cases. Even if the frequency of these cases is underestimated because mutation analysis is limited to the coding region of the genes examined, other unknown genes could be involved in this disease. Consistent with this hypothesis is the fact that linkage analysis has made it possible to exclude the role of either *TITF1*, *PAX8*, *FOXE1* or *TSHR* in a group of families in which at least two members are affected by TD [28].

(b) The mode of transmission of the phenotype could be different between humans and mice. This is the case of *Pax8*$^{+/-}$ mice which display an apparently normal phenotype and *Titf1*$^{+/-}$ mice which present a slightly decreased coordination and mild hyperthyrotropinemia. In contrast, humans carrying heterozygous mutations in either *TITF1* or *PAX8* are affected by TD and in familial cases the mode of inheritance is dominant. This discrepancy could be due to a different sensitivity to gene dosage between mice and men. In humans, a reduced amount of the gene product (haploinsufficiency) causes a clear pathological phenotype even if it is usually less severe than that displayed by homozygous *Titf1* or *Pax8* null mice. A strong possibility is that divergences between humans and mice could be due to the specific genetic background of the strain used in generating the corresponding animal models.

(c) Murine models have proved that Foxe1 is specifically involved in thyroid bud migration. Although in humans an ectopic thyroid is the commonest form of TD, *FOXE1* mutations associated with an ectopic thyroid have not yet been described: all subjects carrying loss-of-function mutations in this gene show an absence of the thyroid. However, since only few patients have been reported, it is premature to conclude that *FOXE1* defects in humans do not cause ectopia.

Interestingly, a recent study reports 3 subjects with an ectopic gland carrying heterozygous missense mutations in the *NKX2-5* gene [29]. This factor could be involved, at least in part, in thyroid development; actually, in *Nkx2-5* null mouse embryos the thyroid bud appears smaller when compared to that of wild-type embryos.

(d) In the familial cases of TD, patients do not display a clear mendelian transmission. There is a higher incidence of subclinical thyroid developmental abnormalities in those families with at least one case of TD. In addition, patients carrying mutations in either *TITF1* or *PAX8* genes are affected by syndromes characterized by incomplete penetrance and variability of the phenotype even among the affected members of the same family. Taken together, these observations strongly suggest that in humans also TD could be a multigenic disease. This working hypothesis has received strong support from a novel multigenic model of TD. Mice double heterozygous for the null allele of *Titf1* and *Pax8* gene display an overt thyroid phenotype (hypoplasia or hemiagenesis of the gland and reduced synthesis of Tg) provided the mutations are present on a specific genetic background, C57BL/6. The same mutations in a different mouse strain (129/Sv) are unable to cause any thyroid defects, indicating that other C57BL/6 specific alleles, in addition to the null mutations in Titf1 and Pax8, are responsible for the emergence of TD. This model establishes that TD can be caused by multiple minor genetic defects [23]. The pool of genes that can be affected is rather large. As table 2 shows, defects in several genes, mostly transcription factors or growth factors, have been demonstrated to impair the development of the thyroid in animal models. Minor defects in these genes or in their targets could also be involved in TD in humans.

(e) Congenital hypothyroidism with TD occurs mostly as a sporadic disease. Furthermore, TD has been found to be discordant in monozygotic twins [30]. These observations strongly suggest that this disease, even though genetic, can be noninheritable. The frequency of these entities is difficult to assess. Noninheritable mechanisms, such as somatic mutations or postzygotic epigenetic events, could be involved in the pathogenesis of human TD [31].

Table 2. Currently available mouse models of TD

Gene symbol	Chromosome	Features of the gene product	Thyroid phenotype of null mice embryos
Hhex	19	transcription factor	athyreosis
Titf1	12	transcription factor	athyreosis
Pax8	2	transcription factor	athyreosis
Foxe1	4	transcription factor	athyreosis or ectopia
Tshr	12	G protein-coupled receptor	defects in functional differentiation
Fgfr2	7	tyrosine kinase receptor	athyreosis
Fgf10	13	peptide growth factor	athyreosis
Nkx2-5	17	transcription factor	thyroid bud hypoplasia
Hoxa3	6	transcription factor	hypoplasia; persistent ultimobranchial body
Eya 1	1	transcription factor	hypoplasia; persistent ultimobranchial body
Edn-1	13	signaling peptide	hypoplasia; absence of isthmus
Tbx1	16	transcription factor	hypoplasia; impaired lobulation
Pax3	1	transcription factor	hypoplasia
Shh	5	morphogen	impaired lobulation
Hoxa5	6	transcription factor	defects in functional differentiation

References

1 Lazzaro D, Price M, De Felice M, Di Lauro R: The transcription factor TTF-1 is expressed at the onset of thyroid and lung morphogenesis and in restricted regions of the foetal brain. Development 1991;113:1093–1104.

2 De Felice M, Di Lauro R: Thyroid development and its disorders: genetics and molecular mechanisms. Endocr Rev 2004;25:722–746.

3 Thomas PQ, Brown A, Beddington R: Hex: a homeobox gene revealing peri-implantation asymmetry in the mouse embryo and an early transient marker of endothelial cell precursors. Development 1998;125:85–95.

4 Zannini MS, Francis-Lang H, Plachov D, Di Lauro R: Pax-8, a paired domain-containing protein, binds to a sequence overlapping the recognition site of a homeodomain and activates transcription from two thyroid-specific promoters. Mol Cell Biol 1992;12:4230–4241.

5 Grapin-Botton A, Melton M: Endoderm development: from patterning to organogenesis. Trends Genet 2000;16:124–130.

6 Parlato R, Rosica A, Rodriguez-Mallon A, Affuso A, Postiglione MP, Arra C, Mansouri A, Kimura S, Di Lauro R, De Felice M: An integrated regulatory network controlling survival and migration in thyroid organogenesis. Dev Biol 2004;276:464–475.

7 Fagman H, Andersson L, Nilsson M: The developing mouse thyroid: embryonic vessel contacts and parenchymal growth pattern during specification, budding, migration, and lobulation. Dev Dyn 2006;235:444–455.

8 Trueba SS, Auge J, Mattei G, Etchevers H, Martinovic J, Czernichow P, Vekemans M, Polak M, Attie-Bitach T: PAX8, TITF1 and FOXE1 gene expression patterns during human development: new insights into human thyroid development and thyroid dysgenesis associated malformations. J Clin Endocrinol Metab 2004;90:455–462.

9 Martinez Barbera JP, Clements M, Thomas P, Rodriguez T, Meloy D, Kioussis D, Beddington RS: The homeobox gene Hex is required in definitive endodermal tissues for normal forebrain, liver and thyroid formation. Development 2000;127:2433–2445.

10 Mansouri A, Chowdhury K, Gruss P: Follicular cells of the thyroid gland require Pax8 gene function. Nat Genet 1998;19:87–90.
11 Kimura S, Hara Y, Pineau T, Fernandez-Salguero P, Fox CH, Ward JM, Gonzalez FJ: The T/ebp null mouse: thyroid-specific enhancer-binding protein is essential for the organogenesis of the thyroid, lung, ventral forebrain, and pituitary. Genes Dev 1996;10:60–69.
12 Kusakabe T, Hoshi N, Kimura S: Origin of the ultimobranchial body cyst: T/ebp/Nkx2.1 expression is required for development and fusion of the ultimobranchial body to the thyroid. Dev Dyn 2006;235:1300–1309.
13 Pohlenz J, Dumitrescu A, Zundel D, Martine U, Schonberger W, Koo E, Weiss RE, Cohen RN, Kimura S, Refetoff S: Partial deficiency of thyroid transcription factor 1 produces predominantly neurological defects in humans and mice. J Clin Invest 2002;109:469–473.
14 Revest JM, Spencer-Dene B, Kerr K, De Moerlooze L, Rosewell I, Dickson C: Fibroblast growth factor receptor 2-IIIb acts upstream of Shh and Fgf4 and is required for limb bud maintenance but not for the induction of Fgf8, Fgf10, Msx1, or Bmp4. Dev Biol 2001;231:47–62.
15 Krude H, Schutz B, Biebermann H, von Moers A, Schnabel D, Neitzel H, Tonnies H, Weise D, Lafferty A, Schwarz S, De Felice M, von Deimling A, van Landeghem F, Di Lauro R, Gruters A: Choreoathetosis, hypothyroidism, and pulmonary alterations due to human NKX2-1 haploinsufficiency. J Clin Invest 2002;109:475–480.
16 Plachov D, Chowdhury K, Walther C, Simon D, Guenet JL, Gruss P: Pax8, a murine paired box gene expressed in the developing excretory system and thyroid gland. Development 1990;110:643–651.
17 Macchia PE, Lapi P, Krude H, Pirro MT, Missero C, Chiovato L, Souabni A, Baserga M, Tassi V, Pinchera A, Fenzi G, Gruters A, Busslinger M, Di Lauro R: PAX8 mutations associated with congenital hypothyroidism caused by thyroid dysgenesis. Nat Genet 1998;19:83–86.
18 Dathan N, Parlato R, Rosica A, De Felice M, Di Lauro R: Distribution of the titf2/foxe1 gene product is consistent with an important role in the development of foregut endoderm, palate, and hair. Dev Dyn 2002;224:450–456.
19 De Felice M, Ovitt C, Biffali E, Rodriguez-Mallon A, Arra C, Anastassiadis K, Macchia PE, Mattei MG, Mariano A, Schoeler H, Macchia V, Di Lauro R: A mouse model for hereditary thyroid dysgenesis and cleft palate. Nat Genet 1998;19:395–398.
20 Fagman H, Grande M, Edsbagge J, Semb H, Nilsson M: Expression of classical cadherins in thyroid development: maintenance of an epithelial phenotype throughout organogenesis. Endocrinology 2003;144:3618–3624.
21 Clifton-Bligh RJ, Wentworth JM, Heinz P, Crisp MS, John R, Lazarus JH, Ludgate M, Chatterjee V: Mutation of the gene encoding human TTF-2 associated with thyroid agenesis, cleft palate and choanal atresia. Nat Genet 1998;19:399–401.
22 Fagman H, Grande M, Gritli-Linde A, Nilsson M: Genetic deletion of sonic hedgehog causes hemiagenesis and ectopic development of the thyroid in mouse. Am J Pathol 2004;164:1865–1872.
23 Amendola E, De Luca P, Macchia PE, Terracciano D, Rosica A, Chiappetta G, Kimura S, Mansouri A, Affuso A, Arra C, Macchia V, Di Lauro R, De Felice M: A mouse model demonstrates a multigenic origin of congenital hypothyroidism. Endocrinology 2005;146:5038–5047.
24 Postiglione MP, Parlato R, Rodriguez-Mallon A, Rosica A, Mithbaokar P, Maresca M, Marians RC, Davies TF, Zannini MS, De Felice M, Di Lauro R: Role of the thyroid-stimulating hormone receptor signaling in development and differentiation of the thyroid gland. Proc Natl Acad Sci USA 2002;99:15462–15467.
25 Marians RC, Ng L, Blair HC, Unger P, Graves PN, Davies TF: Defining thyrotropin-dependent and -independent steps of thyroid hormone synthesis by using thyrotropin receptor-null mice. Proc Natl Acad Sci USA 2002;99:15776–15781.
26 Refetoff S: Resistance to thyrotropin. J Endocrinol Invest 2003;26:770–779.
27 Ledent C, Dumont JE, Vassart G, Parmentier M: Thyroid expression of an A2 adenosine receptor transgene induces thyroid hyperplasia and hyperthyroidism. EMBO J 1992;11:537–542.
28 Castanet M, Sura Trueba S, Chauty A, Carré A, Heath S, Léger J, Lyonnet S, Czernichow P, Polak M: Linkage and mutational analysis of familial thyroid dysgenesis suggest genetic heterogeneity. Eur J Hum Genet 2005;13:232–239.
29 Dentice M, Cordeddu V, Rosica A, Ferrara AM, Santarpia L, Salvatore D, Chiovato L, Perri A, Moschini L, Fazzini C, Olivieri A, Costa P, Stoppioni V, Baserga M, De Felice M, Sorcini M, Fenzi G,

Di Lauro R, Tartaglia M, Macchia PE: Missense mutation in the transcription factor NKX2-5: a novel molecular event in the pathogenesis of thyroid dysgenesis. J Clin Endocrinol Metab 2006;91: 1428–1433.
30 Perry R, Heinrichs C, Bourdoux P, Khoury K, Szots F, Dussault JH, Vassart G, Van Vliet G: Discordance of monozygotic twins for thyroid dysgenesis: implications for screening and for molecular pathophysiology. J Clin Endocrinol Metab 2002;87:4072–4077.
31 Vassart G, Dumont JE: Thyroid dysgenesis: multigenic or epigenetic, or both. Endocrinology 2005;146:5035–5037.

Prof. Roberto Di Lauro
IRGS, Biogem s.c.a r.l.
Via Camporeale
IT–83031 Ariano Irpino (AV) (Italy)
Tel. +39 0825 881630, Fax +39 0825 881637, E-Mail rdilauro@unina.it

Van Vliet G, Polak M (eds): Thyroid Gland Development and Function.
Endocr Dev. Basel, Karger, 2007, vol 10, pp 15–28

Familial Forms of Thyroid Dysgenesis

Mireille Castanet[a], *Michel Polak*[a], *Juliane Léger*[b]

[a]Paediatric Endocrinology Unit and INSERM U845, Hôpital Necker-Enfants Malades and [b]Paediatric Endocrinology Unit, Hôpital Robert Debré, Paris, France

Abstract

In many instances, the pathophysiology of thyroid dysgenesis (TD) remains as yet unclear and until relatively recently the disorder was usually regarded as occurring in a sporadic manner. However, over the past few years, a small but significant proportion of familial cases has been identified (2%) through the study of subjects with congenital hypothyroidism and more recent work has revealed an even higher proportion of familial TD in both symptomatic or asymptomatic individuals (7.9%). Together, these studies strongly point to a significant genetic component of this disorder. Moreover, detailed observations of members affected by different types of TD in the same family suggest that TD could be an entity with a common underlying mechanism for all the etiological groups. To date, molecular genetic studies have implicated four genes in thyroid development and some mutations have been reported in affected subjects. Three of these encode transcription factors while the forth encodes the thyrotropin hormone receptor. However, their involvement in the general TD population remains questionable, as only a few mutations have been reported so far and as linkage analysis has demonstrated the relevance of other genes. Therefore, further work is required to fully understand the pathophysiology of TD.

Introduction

Thyroid dysgenesis (TD) is the most frequent cause of congenital hypothyroidism (CH; 85% of cases) and results from abnormalities of thyroid gland development including a spectrum of embryogenetic defects such as ectopic thyroid gland, athyreosis and more rarely, thyroid hypoplasia and hemiagenesis [1]. Other variations of thyroid anatomy such as cysts of the thyroglossal duct and additional thyroid tissue have been occasionally described, usually in asymptomatic patients with normal thyroid function [2].

The pathogenesis of TD is as yet not well-understood and until relatively recently the disorder was typically regarded as arising in a sporadic manner.

Over the last 50 years, reports of some familial cases of CH by either athyreosis [3–5] or ectopic gland [6–8] suggested an inherited disease and two recent familial studies have confirmed this hypothesis [1, 9, 10]. In this chapter, we describe current knowledge regarding the familial forms of TD and explore the molecular basis of this disorder.

Thyroid Development

In the human embryo the thyroid gland primordium first appears late in the 4th week in the midline of the floor of the primitive pharynx at a point that is later known as the foramen cecum on the tongue. Thereafter, this initially round cluster of cells begins its migration from the pharyngeal floor through the anterior midline of the neck during which time the cells multiply. At 24–32 days, this median anlage has already become a bilobed structure, but it only reaches its final position at around 48–50 days. At the same time, connection of the median anlage with the ultimobranchial body, developed from the endoderm of the 4th pharyngeal pouch, occurs, resulting in the incorporation of the C cells into the thyroid. At 51 days, the gland exhibits its definitive external form, with an isthmus connecting the two lateral lobes and reaches its final position below the thyroid cartilage by the 7th week of embryonic life. During its descent, the developing thyroid gland retains an attachment to the pharynx by a narrow epithelial stalk known as the thyroglossal duct [11]. By 37 days, this structure that connects the median thyroid anlage with the point of origin of its migration on the floor of the pharynx has generally disappeared [12] and normally the only remnant of the thyroglossal duct is the foramen cecum itself (fig. 1).

Usually, the terminal differentiation of thyroid follicular cells [as evidenced by expression of the genes encoding the TSH receptor (TSHR), the Na/I symporter (NIS), thyroglobulin (Tg) and thyroperoxidase (TPO)] and the formation of follicles occur in the normal embryo only once migration is complete [13].

Different Types of TD

The vast majority of CH patients with TD have a defect in thyroid migration which results in the presence of ectopic thyroid tissue (~65% of CH patients). This arrest in the normal descent of part or all of the thyroid gland can result in a lingual, suprahyoid or infrahyoid location. Most frequently, thyroid tissue is found at the base of the tongue. Note that double ectopic thyroid tissues have also been exceptionally reported [14]. Although usually, ectopically sited

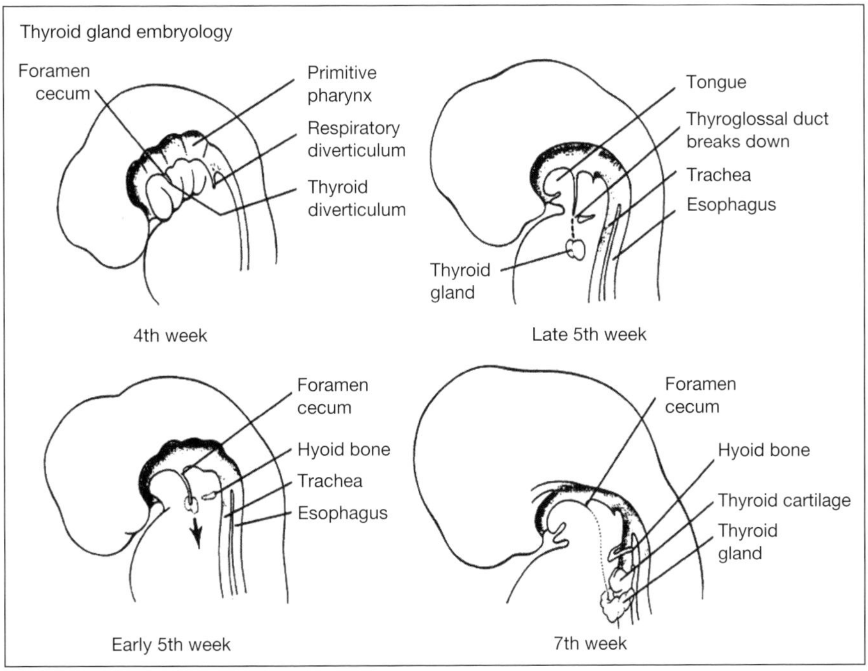

Fig. 1. Thyroid development [11].

thyroid tissues manifest with CH, a proportion of cases are found incidentally in asymptomatic subjects raising the possibility that many are never diagnosed [15]. For example, it has been shown in 200 consecutive necropsies on euthyroid Caucasian individuals that 10% of this population exhibited ectopic lingual thyroid tissue, in addition to a normal thyroid; the size of this ectopic tissue varied from a few acini to a nodule of 1 cm in diameter with males and females equally affected [16]. The fact that some subjects appear to remain asymptomatic suggests that the ectopically sited thyroid tissues could retain normal function. This hypothesis is supported by the finding of uptake of radioactive iodine on scintiscan and of typical colloid-filled follicles under microscopic examination following surgical removal [17]. Moreover, although the thyroxine-producing capacity of patients with ectopically sited thyroid tissue is generally limited, it appears to remain constant, suggesting normal postnatal survival of ectopic cells [18].

The second most common variant of TD (~30%) is the absence of detectable thyroid follicular cells (commonly named athyreosis, although this designation is not entirely correct since these patients, like those with ectopic thyroids, have functional C cells). Whether thyroid follicular cells disappear through apoptosis after initial differentiation [as demonstrated in knockout mice in the chapter by De Felice and Di Lauro, pp. 1–14] or fail to differentiate initially is unknown. Such developmental failure can also affect only one lobe of the gland resulting in hemiagenesis; more rarely it involves the majority, but not all, of the follicular cells resulting in very hypoplastic glands which are normally situated. Note that these small glands are sometimes so hypofunctional that they may be undetectable by nuclear medicine studies: in these cases of apparent athyreosis, careful ultrasonographic evaluation of the neck may reveal a very small thyroid gland of normal shape and in the orthotopic position. This failure to detect any thyroid tissue by scintiscan usually occurs in patients with severe thyrotropin resistance due to complete loss of TSHR function [19, 20]. Therefore, a certain degree of misclassification might exist and to make an accurate diagnosis of athyreosis, it is important to combine a good thyroid scintigraphy with high definition ultrasonography. In addition, undetectable serum thyroglobulin levels in hypothyroid subjects could provide an additional clue in athyreosis cases [21].

Developmental thyroid abnormalities may also include persistence of the thyroglossal duct giving rise to 'thyroglossal duct cysts' which become generally clinically apparent during infancy or childhood. Alternatively, cell residues may remain giving rise to a pyramidal lobe at the distal portion of the duct which remains attached to the thyroid gland, usually to the left lobe [12, 22]. Cysts within the empty thyroid area in CH patients with TD have also been found, suggesting possible cystic degeneration of clusters of thyroid follicular cells that have completed their normal migration, even when the thyroid gland has otherwise remained incompletely descended or has entirely disappeared [14]. Additionally, other adjacent developmental abnormalities have been reported, for example additional thymic tissue within the empty thyroid bed in patients with either ectopic thyroid tissue or athyreosis [23].

Note that the 2 major forms of TD, i.e. ectopic thyroid gland and athyreosis, are often associated with CH, whilst other variations as outlined above are usually found in asymptomatic subjects with normal thyroid function, hence making it difficult to establish their true incidence. Thyroid hemiagenesis and the presence of a pyramidal lobe are indeed usually incidentally discovered in patients with other thyroid disorders when thyroid ultrasound or scintigraphy is performed [22, 24]. So far, only one study has reported a prevalence for thyroid hemiagenesis of 0.2% derived from a systematic ultrasound investigation of normal children [25]. Thyroglossal duct cysts are usually diagnosed through an asymptomatic neck mass, acute infection, chronic inflammation or hemorrhage and account for approximately 70% of congenital neck abnormalities [26].

Familial Forms of CH due to TD

A thorough survey of the literature covering the past 40 years has revealed that almost 20 families with multiple affected individuals with CH due to TD have been described in different countries, suggesting that TD could be a familial disease in some instances [3–8]. Therefore, over the past few years, a French national survey was performed to corroborate this hypothesis. In this survey covering more than 2,500 CH patients with TD diagnosed in France between 1980 and 1998, 67 patients with a positive family history of CH from TD belonging to 32 multiplex families (i.e. with at least two affected members) were referred. These data made it possible to determine a prevalence of familial cases of 2%. Additionally, statistical analysis revealed that, among first-degree relatives, the number of familial cases of CH with TD was significantly higher (by ≥15-fold) than expected by chance alone, indicating that TD revealed by CH had a significant familial component [1, 9].

The pedigrees of the families included in this study are shown in figure 2. Most of the familial cases were identified in families with 2 affected members (n = 30), while 1 family presented with 3, and 1 with 4 affected members. Interestingly, both vertical and horizontal clustering was observed among first-degree relatives (n = 23): 13 families with affected sibs and 10 families with affected parents and offspring. The 10 other families displayed more distant relationships, including 6 families with affected first cousins, one family with affected second cousins and 2 families in which more distantly related members were affected. This high proportion of transmissions from parent to offspring and the possible father to son transmission (n = 2) suggested a possible autosomal dominant mode of inheritance with incomplete penetrance. Nevertheless, these transmission patterns could also suggest genetic heterogeneity [1].

These families also provided important data regarding the etiology of the TD. Indeed, as shown in figure 2, whether familial CH cases were affected by either athyreosis (n = 7 families) or ectopic thyroid gland (n = 12 families), it is noteworthy that in 13 families, athyreosis and ectopic thyroid gland coexisted. Moreover, CH members with eutopic thyroid gland were seen in some families in which at least 2 other members were affected by either athyreosis or ectopic gland. These original findings suggested a common underlying mechanism leading to the defects either in embryogenetic migration, differentiation or growth of the thyroid gland during thyroid organogenesis.

Regarding the sex distribution, while a well-known female preponderance over males was confirmed in isolated CH due to TD, it is interesting to see that the female:male ratio was significantly lower in familial cases (1.4 vs. 2.7; $p < 0.03$). Furthermore, according to the etiological diagnosis of TD, the female:male ratio was significantly reduced in familial compared to isolated

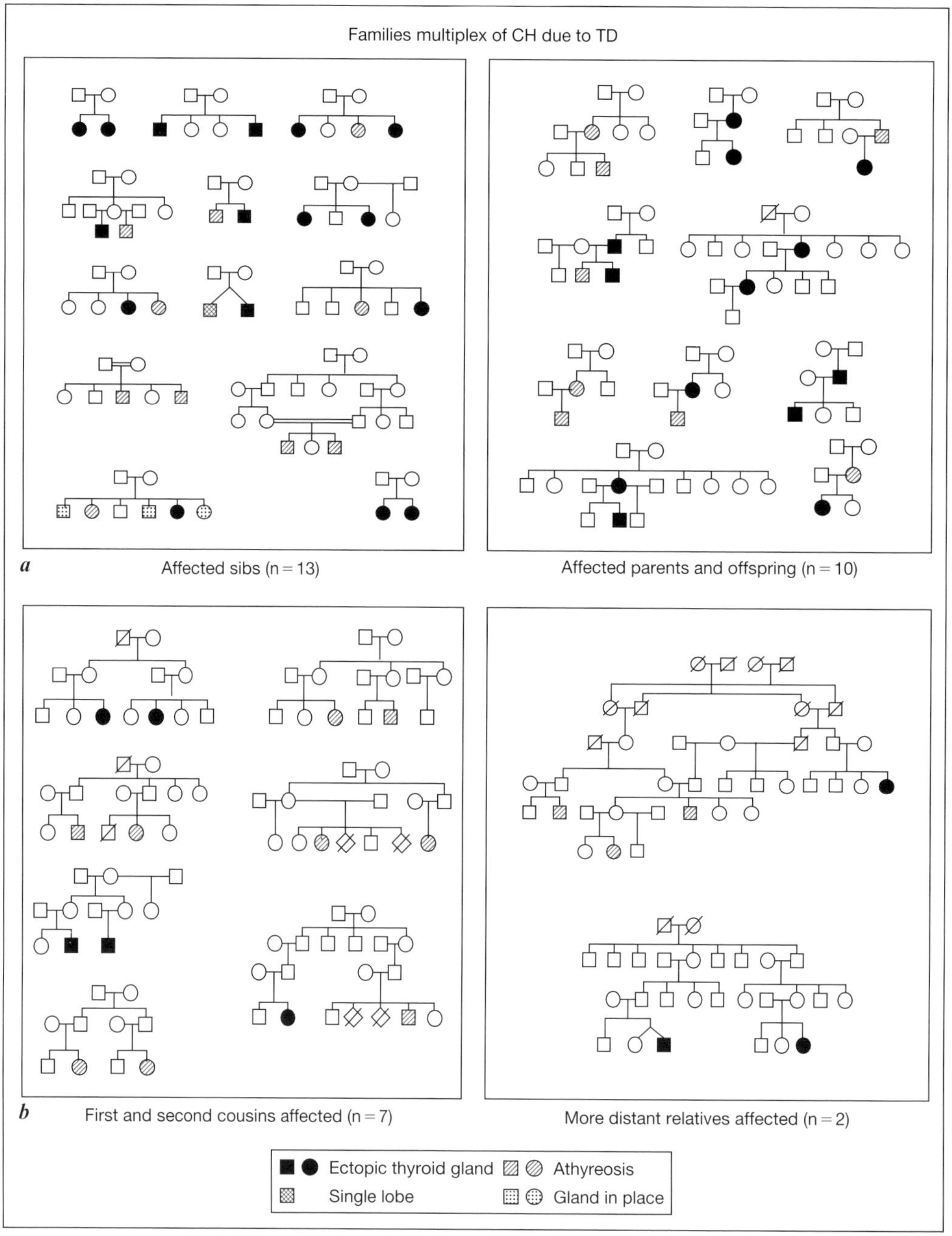

Fig. 2. Pedigrees and phenotypes in the 32 families with CH from TD [1]. ***a*** Families with first-degree relatives affected. ***b*** Families with affected distant relatives. Cross-hatched and black dotted symbols represent CH patients with a TD or an orthotopic thyroid gland respectively.

cases in CH with athyreosis (0.9 vs. 2.7; $p < 0.02$), whereas only a slightly lower proportion (nonsignificant) of females was observed in cases of ectopic thyroid gland (1.9 vs. 2.7). These data could suggest the possible involvement of sex-modified etiological factors in familial cases, particularly in athyreosis.

In conclusion, this family study demonstrated for the first time that TD revealed by CH had a significant familial component, with a potentially common underlying mechanism at least in athyreosis and ectopic thyroid gland. Although common unidentified environmental factors cannot be totally ruled out, this report strongly suggested the existence of genetic factors contributing to the risk of TD that might be controlled by sex. Nevertheless, marked clinical variability both within and between families could reflect genetic heterogeneity and further genetic studies are required to better elucidate the physiopathology of thyroid defects.

Familial Forms of TD in First-Degree Relatives of Children with CH

On the basis that some thyroid tract abnormalities might be totally asymptomatic and that familial cases have been reported with either major forms (i.e. with CH) [1] or TDA (asymptomatic forms of TD) [27] or both in affected members of the same family [6, 28], a further study was performed to determine whether the prevalence of familial cases of TD might be higher than previously reported. Systematic screening by neck ultrasound and measurement of serum TSH and FT_4 concentrations were performed in all first-degree relatives of 84 CH children with TD. Moreover, when ectopic or additional thyroid tissue was found by ultrasound, radioiodine thyroid scanning (radioactive iodide ^{123}I) was performed to identify functional thyroid tissue. Among the 241 screened first-degree relatives of the 84 studied patients, 19 relatives (7.9% of cases) belonging to 18 families were found with asymptomatic TDA, i.e. without any clinical complaints and serum-free thyroid hormone and TSH levels in the normal range. This proportion of affected individuals in the nuclear families of CH patients with TD was significantly higher than that seen in the control population (7.9 vs. 0.9%; $p < 0.001$), and pointed to a familial disorder [10]. As shown in figure 3, the 19 subjects affected by TDA carried a total of 21 detected anomalies as 2 subjects exhibited 2 different disorders, respectively, hemiagenesis and thyroglossal duct cyst (n = 1) and additional thyroid tissue and thyroglossal duct cyst (n = 1). Thyroglossal duct cyst was the main TDA found in 14 subjects (7 males, 7 females) who were the sibs (n = 6) or the parents (n = 8) of 13 CH children with either ectopic thyroid tissue (n = 5), athyreosis (n = 7) or hemiagenesis (n = 1). In all of these 14 patients, the thyroid gland

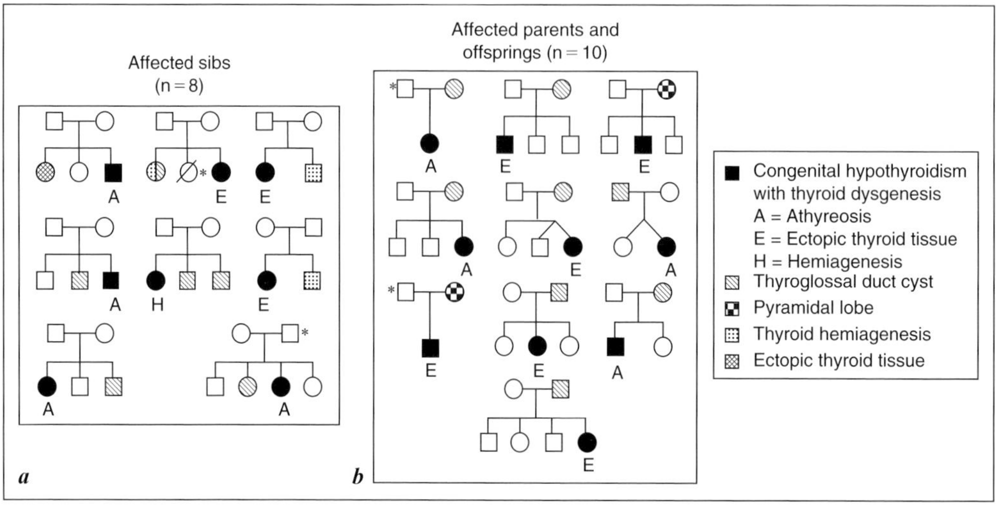

Fig. 3. Pedigrees of the 18 families with TD, both with CH and asymptomatic cases [54]. ***a*** Families with affected sibs (n = 8). ***b*** Families with affected parents and offsprings (n = 10).

was normally located. Additional thyroid tissue with the presence of a pyramidal lobe along the left lobe of the normally located thyroid gland was found in 3 mothers of CH children with ectopic thyroid tissue. Thyroid hemiagenesis with the presence of a single well-located lobe (left: n = 2, right: n = 1) was found in 3 sibs (2 males, 1 females) of 3 CH children with ectopic thyroid tissue. Finally, surprisingly, cervical ectopic thyroid gland was found in a healthy clinically and biologically euthyroid sister of 1 CH child with athyreosis without any evidence of thyroid tissue in the normal location determined by ultrasonography. This observation is in accordance with a substantial function of ectopic thyroid tissue.

Regarding the sex ratio, as for the familial cases of athyreosis, an equal proportion of boys and girls were found in TDA again suggesting possible involvement of a sex-modified gene in thyroid development. Note that these data were in accordance with the equal sex ratio reported in a larger series of symptomatic thyroglossal duct cysts [29].

Regarding the pedigrees reported in this study, a similar proportion of affected parents (6.5%) and sibs (10.5%) with asymptomatic TDA was found, which was again in favor of a dominant mode of inheritance. Moreover, this unique sample of families including members affected with either asymptomatic or major forms of TD made a segregation analysis possible, demonstrating a low penetrance of the disease at 21% and a prediction of the occurrence risk after an isolated case of 10.5% for TDA.

This second familial study demonstrated that among first-degree relatives of a CH population with TD, there is an elevated rate of asymptomatic thyroid developmental anomalies. The estimates of prevalence of families with both minor and major forms of TD (21.4% of the investigated families) were much higher than the proportion of families with only affected members with CH due to TD. Moreover, it should be pointed out that this high proportion of familial occurrence may be underestimated as only first-degree relatives were evaluated by systematic screening in this study excluding more distant relatives which cannot be examined extensively.

In conclusion, these two recent familial studies have revealed a relatively high rate of familial TD including asymptomatic individuals (TDA). Although a role of humoral or environmental factors in the thyroid development cannot be totally excluded, these data strongly suggest a genetic component to the disorder. Moreover, the pedigrees in these 2 studies give support to the concept of a common origin for embryogenesis, migration, differentiation or growth of the thyroid gland during organogenesis. Taken together, these new findings of familial TD cases should prompt us to perform systematic screening for thyroid defects in relatives of patients affected by TD with or without CH.

Molecular Mechanisms of Thyroid Developmental Abnormalities

Studies in knockout mice have demonstrated a critical role for several genes in the early events of thyroid organogenesis. To date, four genes have been involved; three of them encode transcription factors [30] while the other encodes the TSHR. Although the three highly conserved transcription factors (>90% homology between mice and humans) are not thyroid-specific, it is noteworthy that simultaneous expression of these three factors is unique to the thyroid follicular cells from the beginning of their differentiation and is maintained throughout thyroid development and into adulthood [31]. In mice, Ttf-1 inactivation leads to absence of thyroid tissue associated with severe defects in the lung and the forebrain, indicating a critical role of this factor in early events of organogenesis [32]. Ttf-2 inactivation has revealed that this factor is required for the downward migration of the thyroid gland as well as for palate closure, with knockout mice showing either athyreosis or ectopic gland associated with cleft palate. Note that this observation supported the previous hypothesis from familial studies that ectopic thyroid gland and athyreosis could have a common underlying mechanism [33]. Pax8 knockout mice demonstrated severe thyroid hypoplasia with complete absence of follicular structures [34] whilst defects in TSH secretion and action are associated with a small orthotopic thyroid gland, indicating that the action of TSH through its receptor is not required for migration but is essential for the proliferation and for the maintenance of the differentiated function of the thyroid follicular cells [35, 36].

That these genetic factors may be involved in the pathogenesis of TD in humans is supported by recent reports which have identified germline mutations of these 4 candidate genes in about 20 patients with TD. Homozygous TTF-2 (or FOXE1) mutations are reported in 2 familial cases of athyreosis associated with cleft palate and choanal atresia in 1 case. This association, known as Bamforth syndrome, is consistent with the spatio-temporal expression of TTF-2 that has been detected in the thyroid gland and in the oropharyngeal epithelium during development [37–39]. Heterozygous TTF-1 (or NKX2.1) mutations produce a predominantly neurological phenotype (choreoathetosis) with possible pulmonary lesions that is associated with generally mild thyroid dysfunction with a normal or hypoplastic thyroid gland [40–42]. Heterozygous mutations of PAX8 have been identified in some families with an isolated thyroid phenotype (consisting of an orthotopic thyroid hypoplasia) that may be associated with cystic lesions and kidney malformations, e.g. unilateral renal agenesis and a left-sided ureteropelvic obstruction [43, 44]. All of these associations are in keeping with experimental evidence that the proteins are expressed in several other developing organs, lung and ventral forebrain for Ttf1 [32] and kidney for Pax8 [39, 45]. Inactivating mutations of the TSHR gene, either homozygous or compound heterozygous, have been observed in a few cases of thyroid hypoplasia [19, 20, 46, 47]. Therefore, taken together, these data argue in favor of a significant genetic contribution in TD, implicating at least these 4 candidate genes. Nevertheless, further genetic studies are needed to establish phenotype-genotype correlations and to permit genetic counseling for this type of disorder.

However, despite intensive research programs throughout the world, abnormalities in these 4 genes have been found in only a small proportion of TD cases, most of them with syndromic phenotypes, suggesting that none of them is a major genetic factor in this disorder and that other genes may be involved. A linkage analysis performed in 19 TD multiplex families supported this view showing the exclusion of the 4 candidate genes in 5 families that demonstrated for the first time the relevance of other genes [30]. On the basis of function and spatiotemporal expression, several other genes may be involved in thyroid development. For example, the Nkx2.5 gene (or CSX gene in humans), the HOXB3 or HOXA3 genes, the divergent homeobox gene HEX, the hepatocyte nuclear factor HNF3 gene, the GATA6 gene or the eyes absent gene (EYA1) could be relevant candidate genes since they are expressed early during embryogenesis of the thyroid gland and impairment of their function may be responsible for TD. In addition, some of these genes may interact with the previously described candidate genes. However, none of these genes is entirely thyroid specific and their inactivation in mice usually leads to extrathyroidal malformations [48–51] that are not usually found in the TD patients. However, reports of a significantly higher incidence of extrathyroid congenital abnormalities in CH patients than in the general population (respectively, 9 vs. 2.5%) could suggest the implication of genes interacting in the development of several organs [1]. Therefore, careful studies of the thyroid and extrathyroid phenotype in humans and in mice must be performed to identify further relevant candidate genes. Recently, as a higher prevalence of congenital heart disease has been documented in children with CH than in the general population, Dentice et al. [52] investigated the Nkx2.5 gene and found 4 heterozygous mutations in TD patients, suggesting the relevance of this gene to thyroid development. Furthermore, the identification of additional candidate genes controlling early events in thyroid organogenesis in humans and acting upstream of NKX2.1, FOXE1 or PAX8 would be very helpful. Indeed, it is noteworthy that the initial differentiation of thyroid follicular cells on the floor of the pharynx is normal in mice with homozygous deletion of these three transcription factors. In addition, performing the cloning of tissue-specific genes and/or a genomewide screening in a significant number of familial cases of TD could be interesting strategies.

In conclusion, recent data have revealed for the first time the familial character of TD, strongly suggesting a genetically determined disorder. Moreover, many arguments are in favor of a possible polygenic basis for thyroid defects with the involvement of at least 4 genes. However, some of genes involved remain elusive. Additionally, the role of environmental effects and/or the implication of a noninheritable postzygotic event cannot currently be excluded as illustrated by discordance for CH from TD in monozygotic twins [53]. Therefore, accumulating evidence supports the view that the genetics of TD are complex, possibly with a polygenic/multifactorial basis, and genetic susceptibility to TD could lack a simple mendelian pattern of inheritance. Accordingly, further studies are needed to enhance our understanding of the pathophysiology of TD.

Acknowledgment

The authors would like to thank Dr. C. Garel (Hôpital Robert Debré, Paris) for re-reading thyroid ultrasound scans in the familial studies and all the families and the pediatricians taking part in these studies.

This work was supported by grants from Novo Nordisk and Evian, and from the Fondation pour la Recherche Médicale. M.C. was awarded the Young Investigator Award at the ESPE (European Society for Paediatric Endocrinology) Meeting 2006 in Rotterdam, The Netherlands.

References

1 Castanet M, Polak M, Bonaïti-Pellié C, Lyonnet S, Czernichow P, Léger J: Nineteen years of national screening for congenital hypothyroidism: familial cases with thyroid dysgenesis suggest the involvement of genetic factors. J Clin Endocrinol Metab 2001;86:2009.

2 Kaplan EL, Shukla M, Hara H, Ito K: Developmental abnormalities of the thyroid; in De Groot LG (ed): Endocrinology. Philadelphia, Saunders, 1994, pp 893–899.

3 Ainger LE, Kelley VC: Familial athyreotic cretinism: report of 3 cases. J Clin Endocrinol Metab 1955;15:469.

4 Bamforth JS, Hugues IA, Lazarus JH, Weaver CM, Harper PS: Congenital hypothyroidism, spiky hair, and cleft palate. J Med Genet 1989;26:49.

5 Stäger J, Froesch ER: Congenital familial thyroid aplasia. Acta Endocrinol (Copenh) 1981;96:188.

6 Orti E, Castells S, Qazi Q, Inamdar S: Familial thyroid disease: lingual thyroid in two siblings and hypoplasia of a thyroid lobe in a third. J Pediatr 1971;78:675.

7 Kaplan M, Kauli R, Raviv U, Lubin E, Laron Z: Hypothyroidism due to ectopy in siblings. Am J Dis Child 1977;131:1264.

8 Defoer FY, Malher C: Lingual thyroid in two natural brothers. J Clin Invest 1990;13:65.

9 Castanet M, Lyonnet S, Bonaïti-Pellié C, Polak M, Czernichow P, Léger J: Familial forms of thyroid dysgenesis among infants with congenital hypothyroidism. N Engl J Med 2000;343:441.

10 Leger J, Marinovic D, Garel C, Polak M, Czernichow P: High prevalence of thyroid developmental anomalies in first degree relatives of children with congenital hypothyroidism. J Clin Endocrinol Metab 2002;87:575–580.

11 O'Rahilly R: The timing and sequence of events in the development of the human endocrine system during the embryonic period proper. Anat Embryol (Berl) 1983;166:439.

12 Sprinzl GM, Koebke J, Wimmers-Klick J, Eckel HE, Thumfart WF: Morphology of the human thyroglossal tract: a histologic and macroscopic study in infants and children. Ann Otol Rhinol Laryngol 2000;109:1135.

13 Macchia PE: Recent advances in understanding the molecular basis of primary congenital hypothyroidism. Mol Med Today 2000;6:36.

14 Marinovic D, Garel C, Czernichow P, Léger J: Additional phenotypic abnormalities with presence of cysts within the empty area in patients with congenital hypothyroidism with thyroid dysgenesis. J Clin Endocrinol Metab 2003;88:1212–1216.

15 Gillis D, Brnjac L, Perlman K, Sochett EB, Daneman D: Frequency and characteristics of lingual thyroid not detected by screening. J Pediatr Endocrinol Metab 1998;11:229.

16 Sauk JJ Jr: Ectopic lingual thyroid. J Pathol 1970;102:239.

17 Gallo A, Leonetti F, Torri E, Manciocco V, Simonelli M, DeVincentiis M: Ectopic lingual thyroid as unusual cause of severe dysphagia. Dysphagia 2001;16:220.

18 Grant DB, Hulse JA, Jackson DB, Leung SP, Ng WK: Ectopic thyroid: residual function after withdrawal of treatment in infancy and later childhood. Acta Paediatr Scand 1989;78:889.

19 Abramowicz MJ, Duprez L, Parma J, Vassart G, Heinrichs C: Familial congenital hypothyroidism due to inactivating mutation of the thyrotropin receptor causing profound hypoplasia of the thyroid gland. J Clin Invest 1997;99:3018.

20 Gagne N, Parma J, Deal C, Vassart G, Van Vliet G: Apparent congenital athyreosis contrasting with normal plasma thyroglobulin levels and associated with inactivating mutations in the thyrotropin receptor gene: are athyreosis and ectopic thyroid distinct entities? J Clin Endocrinol Metab 1998;83:1771.

21 Czernichow P, Schlumberger M, Pomarede R, Fragu P: Plasma thyroglobulin measurements help determine the type of thyroid defect in congenital hypothyroidism. J Clin Endocrinol Metab 1983;56:242.

22 Siraj QH, Aleem N, Inam-Ur-Rehman A, Qaisar S, Ahmad M: The pyramidal lobe: a scintigraphic assessment. Nucl Med Commun 1989;10:685.

23 Bubuteishvili L, Garel C, Czernichow P, Leger J: Thyroid abnormalities by ultrasonography in neonates with congenital hypothyroidism. J Pediatr 2003;143:759.

24 McHenry CR, Walfish PG, Rosen IB, Lawrence AM, Paloyan E: Congenital thyroid hemiagenesis. Am Surg 1995;61:634.
25 Shabana W, Delange F, Freson M, Osteaux M, de Schepper J: Prevalence of thyroid hemiagenesis ultrasound screening in normal children. Eur J Pediatr 2000;159:456.
26 Rapidis AD, Economidis J, Goumas PD, Langdon JD, Skordalakis A, Tzortzatou F, Anagnostopoulos D, Matsaniotis N: Tumours of the head and neck in children. A clinico-pathological analysis of 1,007 cases. J Craniomaxillofac Surg 1988;16:279.
27 Issa MM, deVries P: Familial occurrence of thyroglossal duct cyst. J Pediatr Surg 1991;26:30.
28 Rosenberg T, Gilboa Y: Familial thyroid ectopy and hemiagenesis. Arch Dis Child 1980;55:639.
29 Solomon JR, Rangecroft L: Thyroglossal-duct lesions in childhood. J Pediatr Surg 1984;19:555.
30 Castanet M, Sura-Trueba S, Chauty A, Carre A, de Roux N, Heath S, Leger J, Lyonnet S, Czernichow P, Polak M: Linkage and mutational analysis of familial thyroid dysgenesis demonstrate genetic heterogeneity implicating novel genes. Eur J Hum Genet 2005;13:232.
31 Damante G, Tell G, Di Lauro R: A unique combination of transcription factors controls differentiation of thyroid cells. Prog Nucleic Acid Res Mol Biol 2001;66:307.
32 Kimura S, Hara Y, Pineau T, Fernandez-Salguero P, Fox CH, Ward JM, Gonzalez FJ: The T/ebp null mouse: thyroid-specific enhancer-binding protein is essential for the organogenesis of the thyroid, lung, and ventral forebrain. Genes Dev 1996;10:60.
33 De Felice M, Ovitt C, Biffali E, Rodriguez-Mallon A, Arra C, Anastassiadis K, Macchia PE, Mattei MG, Mariano A, Schöler H, Macchia V, Di Lauro R: A mouse model for hereditary thyroid dysgenesis and cleft palate. Nat Genet 1998;19:399.
34 Mansouri A, Chowdhury K, Gruss P: Follicular cells of the thyroid gland require Pax8 gene function. Nat Genet 1998;19:87.
35 Postiglione MP, Parlato R, Rodriguez-Mallon A, Rosica A, Mithbaokar P, Maresca M, Marians RC, Davies TF, Zannini MS, De Felice M, Di Lauro R: Role of the thyroid-stimulating hormone receptor signaling in development and differentiation of the thyroid gland. Proc Natl Acad Sci USA 2002;99:15462.
36 Beamer WJ, Eicher EM, Maltais LJ, Southard JL: Inherited primary hypothyroidism in mice. Science 1981;212:61.
37 Clifton-Bligh RJ, Wentworth JM, Heinz P, Crisp MS, John R, Lazarus JH, Ludgate M, Chatterjee K: Mutation of the gene encoding human TTF-2 associated with thyroid agenesis, cleft palate and choanal atresia. Nat Genet 1998;19:399.
38 Castanet M, Park SM, Smith A, Bost M, Léger J, Lyonnet S, Pelet A, Czernichow P, Chatterjee K, Polak M: A novel loss-of-function mutation in TTF-2 is associated with congenital hypothyroidism, thyroid agenesis, and cleft palate. Hum Mol Genet 2002;11:2051.
39 Trueba SS, Auge J, Mattei G, Etchevers H, Martinovic J, Czernichow P, Vekemans M, Polak M, Attie-Bitach T: PAX8, TITF1, and FOXE1 gene expression patterns during human development: new insights into human thyroid development and thyroid dysgenesis-associated malformations. J Clin Endocrinol Metab 2005;90:455.
40 Krude H, Schuetz B, Biebermann H, von Moers A, Schnabel D, Neitzel H, Tönnies H, Weise D, Lafferty A, Schwarz S, DeFelice M, von Deimling A, van Landeghem F, DiLauro R, Grüters A: Choreoathetosis, hypothyroidism, and pulmonary alterations due to human NKX2.1 haploinsufficiency. J Clin Invest 2002;109:475.
41 Pohlenz J, Dumitrescu A, Zundel D, Martiné U, Schönberger W, Koo E, Weiss RE, Cohen RN, Kimura S, Refetoff S: Partial deficiency of thyroid transcription factor 1 produces predominantly neurological defects in humans and mice. J Clin Invest 2002;109:469.
42 Simon-Carré A, Szinnai G, Castanet M, Barat P, Delrue MA, Moutard ML, Raynaud C, Romana S, Sarda P, Ythier H, Léger J, Polak M: Clinical and molecular aspects of five TITF1 gene anomalies (abstract). Horm Res 2006;65(suppl 4):CF2-101.
43 Macchia PE, Lapi P, Krude H, Pirro MT, Missero C, Chiovato L, Souabni A, Baserga M, Tassi V, Pinchera A, Fenzi G, Grüters A, Busslinger M, Di Lauro R: Pax8 mutations associated with congenital hypothyroidism caused by thyroid dysgenesis. Nat Genet 1998;19:83.
44 Meeus L, Gilbert B, Rydlewski C, Parma J, Roussie AL, Abramowicz M, Vilain C, Christophe D, Costagliola S, Vassart G: Characterization of a novel loss of function mutation of PAX8 in a familial

case of congenital hypothyroidism with in-place, normal-sized thyroid. J Clin Endocrinol Metab 2004;89:4285.
45 Plachov D, Chowdhury K, Walther C, Simon D, Guenet JL, Gruss P: Pax8, a murine paired box gene expressed in the developing excretory system and thyroid gland. Development 1990;110:643.
46 Bretones P, Duprez L, Parma J, David M, Vassart G, Rodien P: A familial case of congenital hypothyroidism caused by a homozygous mutation of the thyrotropin receptor gene. Thyroid 2002;11:977.
47 Park SM, Clifton-Bligh RJ, Betts P, Chatterjee VK: Congenital hypothyroidism and apparent athyreosis with compound heterozygosity or compensated hypothyroidism with probable hemizygosity for inactivating mutations of the TSH receptor. Clin Endocrinol (Oxf) 2004;60:220.
48 Lints TJ, Parsons LM, Hartley L, Lyons I, Harvey RP: Nkx2.5: A novel murine homeobox gene expressed in early heart progenitor cells and their myogenic descendants. Development 1993; 119:419.
49 Martinez Barbera JP, Clements M, Thomas P, Rodriguez T, Meloy D, Kioussis D, Beddington RSP: The homeobox gene Hex is required in definitive endodermal tissues for normal forebrain, liver and thyroid formation. Development 2000;127:2433.
50 Xu PX, Zheng W, Laclef C, Maire P, Maas RL, Peters H, Xu X: Eya1 is required for the morphogenesis of mammalian thymus, parathyroid and thyroid. Development 2002;129:3033.
51 Ang SL, Rossant J: HNF3b is essential for node and notochord formation in mouse development. Cell 1994;78:561.
52 Dentice M, Cordeddu V, Rosica A, Ferrara AM, Santarpia L, Salvatore D, Chiovato L, Perri A, Moschini L, Fazzini C, Olivieri A, Costa P, Stoppioni V, Baserga M, De Felice M, Sorcini M, Fenzi G, Di Lauro R, Tartaglia M, Macchia PE: Missense mutation in the transcription factor NKX2–5: a novel molecular event in the pathogenesis of thyroid dysgenesis. J Clin Endocrinol Metab 2006;91:1428.
53 Perry R, Heinrichs C, Bourdoux P, Khoury K, Szotz F, Dussault JH, Vassart G, Van Vliet G: Discordance of monozygotic twins for thyroid dysgenesis: implications for screening and for molecular pathophysiology. J Clin Endocrinol Metab 2002;87:4072.

Juliane Léger, MD
Paediatric Endocrinology Unit, Hôpital Robert Debré
48 boulevard Sérurier
FR–75019 Paris (France)
Tel. +33 1 4003 2354, Fax +33 1 4040 2429, E-Mail juliane.leger@rdb.ap-hop-paris.fr

Van Vliet G, Polak M (eds): Thyroid Gland Development and Function.
Endocr Dev. Basel, Karger, 2007, vol 10, pp 29–42

Possible Non-Mendelian Mechanisms of Thyroid Dysgenesis

Johnny Deladoëy[a], *Gilbert Vassart*[b], *Guy Van Vliet*[a]

[a]Endocrinology Service and Research Center, Department of Pediatrics, Sainte-Justine Hospital, University of Montreal, Montreal, Que., Canada; [b]Genetics Service, Erasme Hospital and Institute of Interdisciplinary Research (IRIBHM), Free University of Brussels (U.L.B.), Brussels, Belgium

Abstract

Most research on the molecular mechanisms of thyroid dysgenesis over the past decade has focussed on the Mendelian mechanisms that may account for the few (<5%) cases in which there is an affected relative. This chapter first reviews methodological issues in the imaging techniques used to classify thyroid dysgenesis into its various forms (ectopic thyroid, agenesis, orthotopic hypoplasia and hemiagenesis). It then reviews the evidence that non-Mendelian mechanisms must be involved in the vast majority of cases of this disease, for which the percentage of sporadic cases and of discordance between monozygotic twins exceeds 95%. Among the mechanisms reviewed are early somatic mutations and epigenetic changes in genes involved in thyroid development such as the thyroid transcription factors TTF-1, TTF-2 and PAX-8. The possible role of extrathyroid genes involved in the control of migration of the median thyroid bud during embryogenesis, such as adhesion molecules, and of vascular factors involved in the stabilization of the bilobed structure of the thyroid is also discussed.

Introduction

While considerable progress has been made over the past decade in our understanding of single gene disorders affecting the thyroid axis, the molecular mechanisms underlying defects in thyroid development represent one of the remaining enigmas in the pathophysiology of thyroid diseases [1]. Yet, these defects collectively account for about 80–85% of cases of congenital hypothyroidism, which affects 1 in 3,000 newborns [2].

There are several arguments against a simple Mendelian basis for congenital hypothyroidism from thyroid dysgenesis (CHTD): (1) it affects females more often than males, and this female preponderance is especially pronounced for defects in thyroid migration (resulting in thyroid ectopy), which affects 3 females for 1 male [3]; this sex ratio is not compatible with simple dominant or recessive mechanisms; (2) it is a sporadic disease in more than 95% of the cases diagnosed by neonatal screening [2], and (3) the corresponding figure for the rate of discordance of monozygotic twins is very similar to that of sporadicity (92%) [4]. These clinical observations explain the low yield of the search for germline mutations in genes encoding transcription factors known to be involved in thyroid development such as TTF-1, TTF-2 and PAX-8 in patients with CHTD (about 20 of 500 tested patients) [5, 6] [reviewed in the chapter by Castanet et al., pp. 15–28].

In spite of the observations described above, the field has remained dominated by the Mendelian line of thought for the past 10 years: for instance, it has been argued that the condition could in fact be dominant but that, until the advent of biochemical screening of neonates, the affected subjects were too mentally deficient to reproduce, leading to the apparent sporadicity [7]. Yet, the mean IQ scores in congenital hypothyroidism patients before screening were about 80, and whether this degree of loss of intellectual potential would be sufficient to affect reproductive fitness is questionable. More recently, a multigenic model has been proposed based on studies of mice of different strains that are double heterozygotes for Pax-8 and Ttf-1 [8]. Genes other than those encoding the three thyroid transcription factors mentioned above may be involved in some families [9]. In this respect, it is worth mentioning that thyroid migration may be a passive phenomenon [10], although this is controversial [11], so that genes expressed by the mesenchyme surrounding the thyroid during its migration should be considered as possible candidates for thyroid ectopy. By way of analogy, X-linked Kallman syndrome is a condition in which the migration of endocrine cells (gonadotropin-releasing hormone-secreting neurons) is defective due to a loss of function of a protein extrinsic to these cells [12]. Lastly, recent evidence for the role of embryonic vessels in thyroid organogenesis [13, 14] suggests that vascular factors should be added to the list of candidate genes.

However, it is clear that mechanisms other than Mendelian inheritance need to be considered to explain the thyroid developmental defect in the vast majority of patients with CHTD. Among these, early somatic mutations or epigenetic modifications have been proposed 10 years ago [15] but have not been adequately explored. Before discussing these two possible mechanisms, a brief review of the different types of thyroid dysgenesis in humans, and of how to distinguish them, is presented.

How to Define and Classify Patients with Thyroid Dysgenesis?

The term 'dysgenesis', most often used for the gonads, means 'defective or abnormal development of an organ'. It is often, but not always, associated with abnormal function of that organ. Thyroid dysgenesis usually encompasses four variants, two common and two very uncommon ones. Whether these variants are part of a spectrum or are distinct entities is unclear at present [16].

(1) Ectopic thyroid (75–80% of cases): In these cases, the only thyroid tissue present is either lingual (and may occasionally be visible) or sublingual (in which case it can only be reliably demonstrated by nuclear medicine scintigraphy; fig. 1). It is assumed that this results from a defect in migration. Because active migration of the thyroid primordial cells has not been conclusively proven, some have suggested that the term relocalization be used [14] but we prefer to use the 'migration' terminology. A premature arrest in migration should be distinguished from the finding of aberrant thyroid tissue, in addition to a normal orthotopic gland, beyond the normal path of migration (usually in the mediastinum). In addition to their abnormal location, ectopic (sub)lingual thyroids have an abnormal shape, being round rather than bilobulated. In normal human embryos, the thyroid already has lateral expansions at 32 days [17]. The round appearance of ectopic thyroids therefore implies either that migration stopped before 32 days or that the lateral expansions disappeared. The role of the vasculature and of the ultimobranchial bodies in stabilizing the lateral lobes has been suggested by studies in mice [13]. Whether the atrophy of the thyroid arteries seen in humans with lingual thyroids [18] represents a cause or a consequence of the regression of the lateral expansions remains unknown.

(2) Agenesis or athyreosis (15–20% of cases): In this situation, there is no uptake of radioisotope (technetium or iodine) in the thyroid bed, along the normal path of descent or anywhere in the cervical area. Interestingly, up to 50% of such patients have a detectable plasma thyroglobulin concentration and their condition is best described as 'apparent' athyreosis [19]. Whether their thyroglobulin-producing tissue is orthotopic or ectopic cannot be determined using current imaging techniques. In patients with undetectable radioisotope uptake *and* plasma thyroglobulin ('true' athyreosis), it is unknown whether their thyroid follicular cells never differentiated or whether they disappeared after initial differentiation. Along this line, it should be mentioned that apoptosis of thyroid cells after initial differentiation has been shown in Ttf-1$^{-/-}$ embryos [20].

(3) Hypoplasia of a bilobed and orthotopic thyroid (less than 5% of cases): These hypoplastic glands typically take up radioisotopes rather poorly or not at all. The reported cases of heterozygous mutations in PAX8 or of

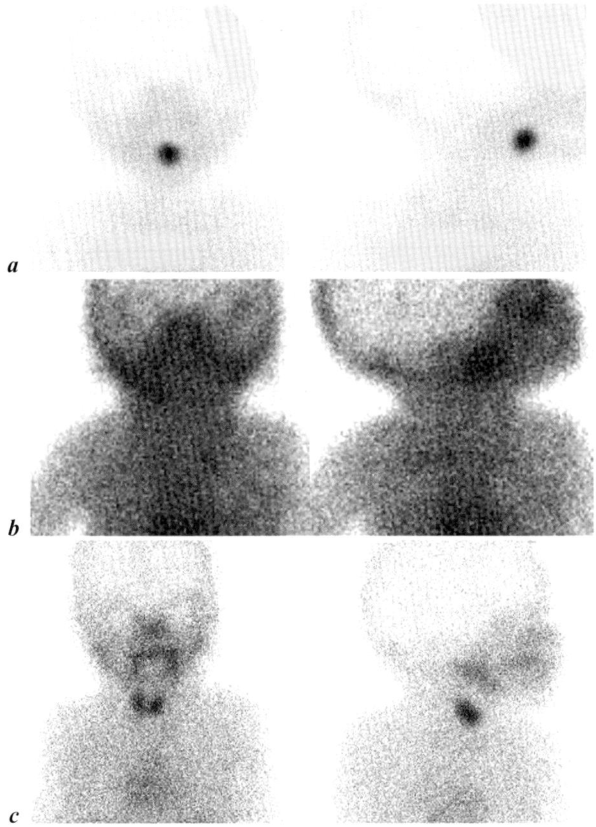

Fig. 1. The two commonest variants of thyroid dysgenesis as documented by sodium pertechnetate scintigraphy. ***a*** Ectopic sublingual thyroid; note the lack of lateral lobes, giving a round appearance, and an absence of uptake in the area where the thyroid normally lies. ***b*** No detectable thyroid tissue; note that the background level has been increased to ascertain that no thyroid tissue could be visualized, which results in some visible uptake in the salivary glands and mediastinum; true athyreosis was confirmed by an undetectable plasma thyroglobulin concentration. ***c*** For comparison, the normal appearance of the thyroid gland in a newborn is shown. Left panels are anterior views, right panels are lateral (right) views.

homozygous or compound heterozygous mutations in the TSH receptor (TSHR) have generally presented with this phenotype [16, 21, 22].

(4) Hemiagenesis: This situation, in which one of the thyroid lobes (usually the left), and sometimes the isthmus, are absent, accounts for less than 1% of cases of congenital hypothyroidism [3]. However, it is observed in 1 in

500 euthyroid children examined by ultrasound [23]. This is because compensatory growth of the remaining tissue permits adequate thyroid hormone production in the vast majority of cases.

The gold standard for distinguishing between these subtypes of thyroid dysgenesis is radioisotope scanning (fig. 1) [reviewed in the chapter by Garel and Léger, pp. 43–61]. The sublingual thyroids can be readily visualized if the child is fed between the administration of the isotope and imaging (to decrease uptake by the salivary glands, in the case of technetium scanning) [24]. However, when a low or absent uptake is seen, there are a number of potential pitfalls that have to be considered (failure to inject or ingest the isotope, presence of maternally derived TSHR blocking antibodies, massive iodine overload, and mutations in the TSHR or in the sodium-iodide symporter). Variations in the technical quality of the isotope scanning likely account for wide differences in the proportion of ectopy and athyreosis between centers.

Ultrasound has been extensively studied, including with color Doppler imaging. It has so far been found to be inferior to isotope scanning in detecting sublingual thyroids [25]. In the area where the thyroid normally lies in cases of proven sublingual thyroid or of true athyreosis, the hyperechogenic structures that can be identified by very experienced radiologists likely represent the ultimobranchial bodies [26].

Given theses caveats, the phenotypic description of patients with a mutation purportedly causing a certain type of thyroid dysgenesis should be carefully scrutinized. That apparent athyreosis or orthotopic thyroid hypoplasia may result from heterozygous PAX8 mutations [21] or from homozygous or compound heterozygous TSHR mutations [16, 22] is well documented. Likewise, well-documented cases illustrate that homozygous TTF-2 mutations lead to true athyreosis [27–29], while mutations in GLI3 lead to apparent athyreosis [30], in both situations with a number of extrathyroid manifestations. However, claims that thyroid tissue is present on ultrasound when plasma thyroglobulin and radioisotope uptake are undetectable [31] should be viewed with great scepticism. So should statements that the thyroid was ectopic without specifying which type of imaging was used [7, 32]. Thus, because no case of thyroid ectopy documented by scintigraphy has been linked to a germline mutation in any transcription factor, this disease entity is the prime category for searching other disease mechanisms, which are discussed below.

Early Somatic Mutations

Postzygotic events, such as an early somatic mutation with a dominant effect leading to a loss of function in a gene important for thyroid development, could theoretically explain a sporadic occurrence and the habitual discordance between monozygotic twins. A conceptual criticism of this hypothesis has been that somatic mutations usually lead to a phenotype because they result in a gain of function, the mutated cell having acquired a competitive advantage over its neighbors such as an increased rate of proliferation or increased function. The best example of such a phenomenon in the field of thyroid diseases is the role of somatic gain-of-function mutations in the pathophysiology of hyperfunctioning thyroid adenomas [33]. If a loss of function occurred in 1 of the about 50 cells when they start forming the median thyroid bud, that cell may die, fail to migrate and/or to undergo terminal differentiation, but all the others will follow their normal developmental program and there would therefore be no phenotype. Thus, the somatic mutation hypothesis implies that the mutational event occurred very early during development, in the cell that is the common ancestor of the cells destined to become thyroid follicular cells. In this respect, it is important to mention that studies of the clonality of the thyroid have revealed that it appears to have a large 'embryonic patch size' (in other words, it is derived from only a few cell clones – as few as two, one for each lobe in one of the specimens reported [34]). Furthermore, examples of postzygotic loss-of-function events leading to a phenotype do exist. These include: (1) the occurrence of 45,X/46,XY [35] or 45,X/46,XX [36] mosaic karyotypes, which have been shown to result in variable features of Turner syndrome; (2) the observation of somatic mosaicism in some individuals with androgen insensitivity syndrome [37, 38], and (3) the report of a somatic mutation in *IRF6* in the affected twin of a monozygotic pair discordant for the Van der Woude and popliteal pterygium syndromes [39].

On the other hand, if migration and terminal differentiation are mutually exclusive phenomena, as suggested by the sequence of events observed during normal thyroid development [reviewed in the chapter by De Felice and Di Lauro, pp. 1–14], a gain-of-function mutation leading to a premature start of the terminal differentiation program could result in a secondary arrest in migration. It should be emphasized that ectopic thyroid glands have undergone complete terminal differentiation, as shown by their capacity to trap and organify iodine and to synthesize thyroid hormone. The hypothyroidism most often observed in patients with ectopic thyroid may be due to a reduced capacity of ectopic tissue to respond adequately to the growth-stimulating effects of TSH, whereas it is fully responsive for activation of function. However, TSH and its receptor are not involved in migration itself [5]. On the basis of results of the perchlorate discharge tests, it has been suggested that some newborns with ectopic thyroids also have impaired iodine organification [40]. However, this is unlikely to play a role in the associated hypothyroidism because hypothyroidism was permanent whereas the 'organification defect' was no longer present after the age of 2 years.

Rather, this suggests that the use of the same normative data to interpret the perchlorate discharge test in newborns and in older children and adults is inappropriate. The histological appearance of ectopic thyroids is essentially normal [41] (our own unpubl. observations), the only abnormality being their round shape and their position.

A practical difficulty in exploring the somatic mutation hypothesis is that it would require that the sequence of candidate genes be compared between DNA extracted from the affected tissue (that is, the ectopic thyroid) and leukocytes from the same individual. Interestingly, Wilkins [42] in 1965 no longer stated, as in previous editions of his textbook, that ectopic thyroids were to be removed. This earlier recommendation had been based on the fear of malignant transformation of ectopic thyroid tissue; the concept that ectopic thyroids would be more prone to malignant degeneration was based on an analogy between 'undescended thyroid' and undescended testes. While there are cases of cancer occurring in an ectopic gland in the literature [43], these likely represent a reporting bias and there is no evidence that malignant transformation is more common in ectopic than in orthotopic thyroids. Also, nowadays most patients with ectopic thyroids are identified in the neonatal period because they have an increased TSH on neonatal screening. Treatment with thyroxine normalizes TSH and prevents proliferation and growth of the ectopic thyroid which therefore will generally never cause obstructive symptoms. Thus, ectopic thyroids are very seldom surgically removed. Furthermore, the very few paraffin-embedded ectopic thyroids that can be retrieved from pathology departments is a poor source of DNA because the DNA that can be extracted is fragmented [44] and therefore not suitable for mutation screening. The somatic mutation hypothesis is also pursued by investigators interested in congenital heart malformations [45]. However, we think that the reports of multiple sequence variants in several genes in the diseased area of the heart and not in the normal area are likely artifactual, due to the use of formalin-embedded specimens.

In spite of these conceptual and practical difficulties, we remain convinced that the somatic mutation hypothesis still deserves consideration. However, a second hypothesis, that epigenetic factors are important for thyroid development, needs to be addressed because it could also account for the predominantly sporadic occurrence of thyroid dysgenesis and for the discordance between monozygotic twins for this malformation. Indeed, Mathis and Nicolas [46] consider this second hypothesis to be more likely: according to their calculations, the size of founder cell populations in the embryo is too low for the frequency of spontaneous somatic mutations to be a major issue in the morphogenesis of individuals and is more consistent with the frequency of errors generated by epigenetic controls such as methylation of nucleotides.

Epigenetic Modifications

The term epigenetic was initially coined to describe 'the interaction of genes with their environment that bring the phenotype into being'. Nowadays, it encompasses the study of the mechanisms that control changes in gene expression that can be transmitted through a somatic cell linage or even through the germ cells but that cannot be explained by changes in DNA sequence [47]. The most widely studied chemical substrate of epigenesis is the methylation of the 5-carbon position of cytosine by a group of enzymes known as methyltransferases. Methyltransferases specifically target cytosines within the CpG dinucleotide, using S-adenosyl-L-methionine as the methyl donor. Methylation of cytosine at the C^5 position of adenine is the only methylation reaction that is known to occur in vertebrates [48]. 5-Methyl cytosines comprise about 1% of the DNA bases in the genome (approximately 75% of the CpG dinucleotides are methylated) and occur in a nonrandom distribution, with a higher frequency in the 5′ end of genes compared to their 3′ end. CpG dinucleotides are grouped into clusters called CpG islands [49]. Relatively recently, the concept of epigenesis has been extended to encompass all mechanisms associated with cellular differentiation. These include histone acetylation and methylation, chromatin modification and control of mRNA expression by noncoding RNAs [47].

DNA methylation is involved in many normal processes including the regulation of gene transcription, the maintenance of chromatin structure in a tightly coiled configuration, the inactivation of the X chromosome, the silencing of 'parasitic' DNA elements and the marking of imprinted genes in which the transcription of the gene occurs in a parent-of-origin-specific fashion [50]. Aberrant DNA methylation of imprinted genes is involved in an increasing number of human diseases [47]. The disease process in which epigenetic mechanisms have been most widely studied is carcinogenesis: indeed, methylation-induced silencing of tumor suppressor genes seems to play an important role in tumor development and hypomethylating agents are being studied in chemotherapy protocols [51].

During development, the establishment of tissue-specific patterns of gene expression is correlated with specific methylation/demethylation patterns of CpG islands associated with gene promoters [52]. As far as the thyroid is concerned, tissue-specific expression of the thyroglobulin gene has been shown to correlate with the unmethylated state of its promoter [53]. Also, GABP (a ubiquitous transcription factor) has been shown to regulate transcription of the TSHR gene in a methylation-dependent manner. Both methylation of specific CpG sites within the TSHR promoter and methylation sensitivity of GABP appear to contribute to the failure of rat FRT thyroid cells to express the endogenous TSHR. In contrast, these CpG sites are completely demethylated and bind GABP in the rat thyroid cells FRTL-5 expressing the TSHR gene [54].

However, most relevant to developmental biology in general and to the problem of thyroid dysgenesis in particular is the observation that the majority of methylation 'errors' probably occur early in postzygotic development. On the other hand, monozygotic twins, while genetically identical, show differences in their patterns of DNA methylation and histone acetylation that increase during their life span [55]. Specifically, in monozygotic twins discordant for either the Beckwith-Wiedemann syndrome or the Silver-Russel syndrome (which are both linked to the imprinted gene cluster at 11p15.5), different methylation profiles in the 11p15.5 region have been shown [56, 57]. In a monozygotic twin pair discordant for a caudal duplication anomaly, the promoter region of the AXIN gene was more methylated in the affected than in the unaffected twin [58]. Monozygotic twinning by itself is associated with an increased rate of both epigenetic alterations *and* of congenital malformations [1]. Whether methylation differences exist in the promoter of thyroid-specific genes between ectopic and normal thyroid or between monozygotic twins discordant for thyroid dysgenesis remains to be determined.

Towards a Unifying Hypothesis Combining Germline and Somatic Changes in a Two-Hit Model

We are still only at the beginning of our understanding of the molecular mechanisms controlling normal and abnormal thyroid gland development in humans. Studies of animal models and of the few human cases with an identified monogenic disorder responsible for a defect in thyroid gland development have yielded important insights into these mechanisms. However, it is important to consider both Mendelian and non-Mendelian mechanisms and to consider not only molecules expressed in thyroid follicular cells but also those expressed in the surrounding parenchyma and in the developing blood vessels that supply the thyroid during its development.

A unifying hypothesis that could account for the observations reviewed above is a two-hit model combining a germline mutation with a somatic mutation or epigenetic difference in threshold-sensitive genes involved in critical developmental steps such as migration of the thyroid anlage [59] (fig. 2). However, this putative germline mutation is unlikely to be in PAX-8, TTF-1 or TTF-2, because systematic screening of relatively large numbers of patients with CHTD for mutations in these genes has yielded negative results [60–63]. Another sporadic congenital endocrine disorder that is much less common than thyroid dysgenesis, focal hyperinsulinism, has been shown to result from such a

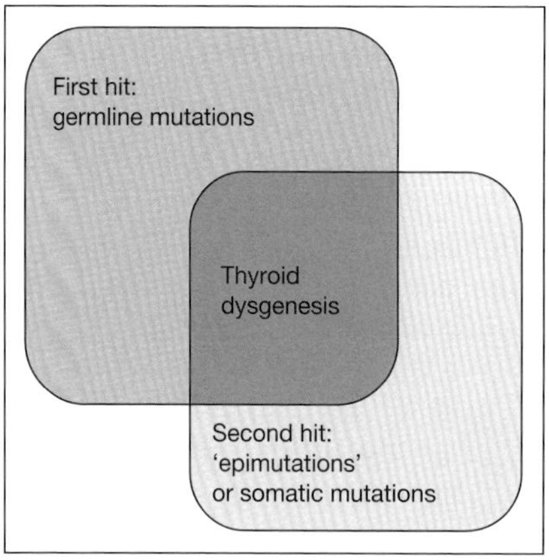

Fig. 2. Two-hit model for thyroid dysgenesis.

two-hit model [64]: in the pancreatic lesions found in these patients, a paternally inherited mutation in the *SUR1* or *KIR6.2* genes is found together with loss of the maternal 11p15 allele (loss of heterozygosity). The loss of heterozygosity is a somatic event restricted to the pancreatic lesion, which explains why focal congenital hyperinsulinism is a sporadic disease. Whether such a model is applicable to thyroid dysgenesis remains to be determined.

Acknowledgments

We would like to thank Dr. Cheri Deal for helpful discussions. Dr. Johnny Deladoëy is supported by the 'Fondation suisse pour les bourses en médecine et biologie' and has received support from the Department of Pediatrics of the University of Montreal and from the Endocrine Fellows Foundation (USA). Research in pediatric thyroid diseases at the Sainte-Justine Hospital is supported by generous donations from Mr. John H. McCall MacBain. Dr. Gilbert Vassart is supported by the Belgian 'Fonds National de la Recherche Scientifique'.

References

1 Vassart G, Dumont JE: Thyroid dysgenesis: multigenic or epigenetic or both? Endocrinology 2005;146:5035–5037.

2 Van Vliet G: Hypothyroidism in infants and children; in Braverman LE, Utiger RD (eds): The Thyroid: a Fundamental and Clinical Text, ed 9. New York, Lippincott Williams & Wilkins, 2005, pp 1029–1047.

3 Devos H, Rodd C, Gagne N, Laframboise R, Van Vliet G: A search for the possible molecular mechanisms of thyroid dysgenesis: sex ratios and associated malformations. J Clin Endocrinol Metab 1999;84:2502–2506.

4 Perry R, Heinrichs C, Bourdoux P, Khoury K, Szots F, Dussault JH, Vassart G, Van Vliet G: Discordance of monozygotic twins for thyroid dysgenesis: implications for screening and for molecular pathophysiology. J Clin Endocrinol Metab 2002;87:4072–4077.

5 Van Vliet G: Development of the thyroid gland: lessons from congenitally hypothyroid mice and men. Clin Genet 2003;63:445–455.
6 De Felice M, Di Lauro R: Thyroid development and its disorders: genetics and molecular mechanisms. Endocr Rev 2004;25:722–746.
7 Macchia PE, Lapi P, Krude H, Pirro MT, Missero C, Chiovato L, Souabni A, Baserga M, Tassi V, Pinchera A, Fenzi G, Gruters A, Busslinger M, Di Lauro R: PAX8 mutations associated with congenital hypothyroidism caused by thyroid dysgenesis. Nat Genet 1998;19:83–86.
8 Amendola E, De Luca P, Macchia PE, Terracciano D, Rosica A, Chiappetta G, Kimura S, Mansouri A, Affuso A, Arra C, Macchia V, Di Lauro R, De Felice M: A mouse model demonstrates a multigenic origin of congenital hypothyroidism. Endocrinology 2005;146:5038–5047.
9 Castanet M, Sura-Trueba S, Chauty A, Carre A, de Roux N, Heath S, Leger J, Lyonnet S, Czernichow P, Polak M: Linkage and mutational analysis of familial thyroid dysgenesis demonstrate genetic heterogeneity implicating novel genes. Eur J Hum Genet 2005;13:232–239.
10 Fagman H, Grande M, Edsbagge J, Semb H, Nilsson M: Expression of classical cadherins in thyroid development: maintenance of an epithelial phenotype throughout organogenesis. Endocrinology 2003;144:3618–3624.
11 Parlato R, Rosica A, Rodriguez-Mallon A, Affuso A, Postiglione MP, Arra C, Mansouri A, Kimura S, Di Lauro R, De Felice M: An integrated regulatory network controlling survival and migration in thyroid organogenesis. Dev Biol 2004;276:464–475.
12 Legouis R, Hardelin JP, Levilliers J, Claverie JM, Compain S, Wunderle V, Millasseau P, Le Paslier D, Cohen D, Caterina D: The candidate gene for the X-linked Kallmann syndrome encodes a protein related to adhesion molecules. Cell 1991;67:423–435.
13 Fagman H, Andersson L, Nilsson M: The developing mouse thyroid: embryonic vessel contacts and parenchymal growth pattern during specification, budding, migration, and lobulation. Dev Dyn 2006;235:444–455.
14 Alt B, Elsalini OA, Schrumpf P, Haufs N, Lawson ND, Schwabe GC, Mundlos S, Gruters A, Krude H, Rohr KB: Arteries define the position of the thyroid gland during its developmental relocalisation. Development 2006;133:3797–3804.
15 Abramowicz MJ, Vassart G, Refetoff S: Probing the cause of thyroid dysgenesis. Thyroid 1997;7:325–326.
16 Gagne N, Parma J, Deal C, Vassart G, Van Vliet G: Apparent congenital athyreosis contrasting with normal plasma thyroglobulin levels and associated with inactivating mutations in the thyrotropin receptor gene: are athyreosis and ectopic thyroid distinct entities? J Clin Endocrinol Metab 1998;83:1771–1775.
17 O'Rahilly R: The timing and sequence of events in the development of the human endocrine system during the embryonic period proper. Anat Embryol (Berl) 1983;166:439–451.
18 Declerck S, Casselman JW, Depondt M, Vandevoorde P: Lingual thyroid imaging. J Belge Radiol 1993;76:241–242.
19 Djemli A, Fillion M, Belgoudi J, Lambert R, Delvin EE, Schneider W, Van Vliet G: Twenty years later: a reevaluation of the contribution of plasma thyroglobulin to the diagnosis of thyroid dysgenesis in infants with congenital hypothyroidism. Clin Biochem 2004;37:818–822.
20 Kimura S, Ward JM, Minoo P: Thyroid-specific enhancer-binding protein/thyroid transcription factor 1 is not required for the initial specification of the thyroid and lung primordia. Biochimie 1999;81:321–327.
21 Meeus L, Gilbert B, Rydlewski C, Parma J, Roussie AL, Abramowicz M, Vilain C, Christophe D, Costagliola S, Vassart G: Characterization of a novel loss of function mutation of PAX8 in a familial case of congenital hypothyroidism with in-place, normal-sized thyroid. J Clin Endocrinol Metab 2004;89:4285–4291.
22 Abramowicz MJ, Duprez L, Parma J, Vassart G, Heinrichs C: Familial congenital hypothyroidism due to inactivating mutation of the thyrotropin receptor causing profound hypoplasia of the thyroid gland. J Clin Invest 1997;99:3018–3024.
23 Shabana W, Delange F, Freson M, Osteaux M, De Schepper J: Prevalence of thyroid hemiagenesis: ultrasound screening in normal children. Eur J Pediatr 2000;159:456–458.
24 Verelst J, Chanoine JP, Delange F: Radionuclide imaging in primary permanent congenital hypothyroidism. Clin Nucl Med 1991;16:652–655.

25 Perry RJ, Maroo S, Maclennan AC, Jones JH, Donaldson MD: Combined ultrasound and isotope scanning is more informative in the diagnosis of congenital hypothyroidism than single scanning. Arch Dis Child 2006;91:972–976.
26 Chanoine JP, Toppet V, Body JJ, Van Vliet G, Lagasse R, Bourdoux P, Spehl M, Delange F: Contribution of thyroid ultrasound and serum calcitonin to the diagnosis of congenital hypothyroidism. J Endocrinol Invest 1990;13:103–109.
27 Bamforth JS, Hughes IA, Lazarus JH, Weaver CM, Harper PS: Congenital hypothyroidism, spiky hair, and cleft palate. J Med Genet 1989;26:49–51.
28 Clifton-Bligh RJ, Wentworth JM, Heinz P, Crisp MS, John R, Lazarus JH, Ludgate M, Chatterjee VK: Mutation of the gene encoding human TTF-2 associated with thyroid agenesis, cleft palate and choanal atresia. Nat Genet 1998;19:399.
29 Castanet M, Park SM, Smith A, Bost M, Leger J, Lyonnet S, Pelet A, Czernichow P, Chatterjee K, Polak M: A novel loss-of-function mutation in TTF-2 is associated with congenital hypothyroidism, thyroid agenesis and cleft palate. Hum Mol Genet 2002;11:2051–2059.
30 Senee V, Chelala C, Duchatelet S, Feng D, Blanc H, Cossec JC, Charon C, Nicolino M, Boileau P, Cavener DR, Bougneres P, Taha D, Julier C: Mutations in GLIS3 are responsible for a rare syndrome with neonatal diabetes mellitus and congenital hypothyroidism. Nat Genet 2006;38: 682–687.
31 Baris I, Arisoy AE, Smith A, Agostini M, Mitchell CS, Park SM, Halefoglu AM, Zengin E, Chatterjee VK, Battaloglu E: A novel missense mutation in human TTF-2 (FKHL15) gene associated with congenital hypothyroidism but not athyreosis. J Clin Endocrinol Metab 2006;91: 4183–4187.
32 Dentice M, Cordeddu V, Rosica A, Ferrara AM, Santarpia L, Salvatore D, Chiovato L, Perri A, Moschini L, Fazzini C, Olivieri A, Costa P, Stoppioni V, Baserga M, De Felice M, Sorcini M, Fenzi G, Di Lauro R, Tartaglia M, Macchia PE: Missense mutation in the transcription factor NKX2-5: a novel molecular event in the pathogenesis of thyroid dysgenesis. J Clin Endocrinol Metab 2006;91:1428–1433.
33 Parma J, Duprez L, Van Sande J, Cochaux P, Gervy C, Mockel J, Dumont J, Vassart G: Somatic mutations in the thyrotropin receptor gene cause hyperfunctioning thyroid adenomas. Nature 1993;365:649–651.
34 Jovanovic L, Delahunt B, McIver B, Eberhardt NL, Grebe SK: Thyroid gland clonality revisited: The embryonal patch size of the normal human thyroid gland is very large, suggesting X-chromosome inactivation tumor clonality studies of thyroid tumors have to be interpreted with caution. J Clin Endocrinol Metab 2003;88:3284–3291.
35 Costa T, Lambert M, Teshima I, Ray PN, Richer CL, Dallaire L: Monozygotic twins with 45,X/46,XY mosaicism discordant for phenotypic sex. Am J Med Genet 1998;75:40–44.
36 Lebl J, Zahradnikova M, Vlasak I, Neuhuber F: Discordant growth pattern and ovarian function in monozygotic twins with 45,X/46,XX mosaicism. Horm Res 2001;55:102–105.
37 Holterhus PM, Bruggenwirth HT, Hiort O, Kleinkauf-Houcken A, Kruse K, Sinnecker GH, Brinkmann AO: Mosaicism due to a somatic mutation of the androgen receptor gene determines phenotype in androgen insensitivity syndrome. J Clin Endocrinol Metab 1997;82: 3584–3589.
38 Kohler B, Lumbroso S, Leger J, Audran F, Grau ES, Kurtz F, Pinto G, Salerno M, Semitcheva T, Czernichow P, Sultan C: Androgen insensitivity syndrome: somatic mosaicism of the androgen receptor in seven families and consequences for sex assignment and genetic counseling. J Clin Endocrinol Metab 2005;90:106–111.
39 Kondo S, Schutte BC, Richardson RJ, Bjork BC, Knight AS, Watanabe Y, Howard E, de Lima RL, Daack-Hirsch S, Sander A, Donald-McGinn DM, Zackai EH, Lammer EJ, Aylsworth AS, Ardinger HH, Lidral AC, Pober BR, Moreno L, Arcos-Burgos M, Valencia C, Houdayer C, Bahuau M, Moretti-Ferreira D, Richieri-Costa A, Dixon MJ, Murray JC: Mutations in IRF6 cause Van der Woude and popliteal pterygium syndromes. Nat Genet 2002;32:285–289.
40 al-Jurayyan NA, el-Desouki MI: Transient iodine organification defect in infants with ectopic thyroid glands. Clin Nucl Med 1997;22:13–16.
41 Gallo A, Leonetti F, Torri E, Manciocco V, Simonelli M, DeVincentiis M: Ectopic lingual thyroid as unusual cause of severe dysphagia. Dysphagia 2001;16:220–223.

42 Wilkins L: The Diagnosis and Treatment of Endocrine Disorders in Childhood and Adolescence. Springfield, Charles C. Thomas, 1965.
43 Perez JS, Munoz M, Naval L, Blasco A, Diaz FJ: Papillary carcinoma arising in lingual thyroid. J Craniomaxillofac Surg 2003;31:179–182.
44 Ando T, Imaizumi M, Graves PN, Unger P, Davies TF: Intrathyroidal fetal microchimerism in Graves' disease. J Clin Endocrinol Metab 2002;87:3315–3320.
45 Reamon-Buettner SM, Borlak J: Somatic mutations in cardiac malformations. J Med Genet 2006;43:e45.
46 Mathis L, Nicolas JF: Cellular patterning of the vertebrate embryo. Trends Genet 2002;18: 627–635.
47 Rakyan VK, Beck S: Epigenetic variation and inheritance in mammals. Curr Opin Genet Dev 2006;16:573–577.
48 Bheemanaik S, Reddy YV, Rao DN: Structure, function and mechanism of exocyclic DNA methyltransferases. Biochem J 2006;399:177–190.
49 Attwood JT, Yung RL, Richardson BC: DNA methylation and the regulation of gene transcription. Cell Mol Life Sci 2002;59:241–257.
50 Robertson KD: DNA methylation and chromatin – unraveling the tangled web. Oncogene 2002;21:5361–5379.
51 Stebbing J, Bower M, Syed N, Smith P, Yu V, Crook T: Epigenetics: an emerging technology in the diagnosis and treatment of cancer. Pharmacogenomics 2006;7:747–757.
52 Klose RJ, Bird AP: Genomic DNA methylation: the mark and its mediators. Trends Biochem Sci 2006;31:89–97.
53 Libert F, Vassart G, Christophe D: Methylation and expression of the human thyroglobulin gene. Biochem Biophys Res Commun 1986;134:1109–1113.
54 Yokomori N, Tawata M, Saito T, Shimura H, Onaya T: Regulation of the rat thyrotropin receptor gene by the methylation-sensitive transcription factor GA-binding protein. Mol Endocrinol 1998;12:1241–1249.
55 Fraga MF, Ballestar E, Paz MF, Ropero S, Setien F, Ballestar ML, Heine-Suner D, Cigudosa JC, Urioste M, Benitez J, Boix-Chornet M, Sanchez-Aguilera A, Ling C, Carlsson E, Poulsen P, Vaag A, Stephan Z, Spector TD, Wu YZ, Plass C, Esteller M: Epigenetic differences arise during the lifetime of monozygotic twins. Proc Natl Acad Sci USA 2005;102:10604–10609.
56 Weksberg R, Shuman C, Caluseriu O, Smith AC, Fei YL, Nishikawa J, Stockley TL, Best L, Chitayat D, Olney A, Ives E, Schneider A, Bestor TH, Li M, Sadowski P, Squire J: Discordant KCNQ1OT1 imprinting in sets of monozygotic twins discordant for Beckwith-Wiedemann syndrome. Hum Mol Genet 2002;11:1317–1325.
57 Gicquel C, Rossignol S, Cabrol S, Houang M, Steunou V, Barbu V, Danton F, Thibaud N, Le Merrer M, Burglen L, Bertrand AM, Netchine I, Le Bouc Y: Epimutation of the telomeric imprinting center region on chromosome 11p15 in Silver-Russell syndrome. Nat Genet 2005;37: 1003–1007.
58 Oates NA, van Vliet J, Duffy DL, Kroes HY, Martin NG, Boomsma DI, Campbell M, Coulthard MG, Whitelaw E, Chong S: Increased DNA methylation at the AXIN1 gene in a monozygotic twin from a pair discordant for a caudal duplication anomaly. Am J Hum Genet 2006;79:155–162.
59 Grasberger H, Vaxillaire M, Pannain S, Beck JC, Mimouni-Bloch A, Vatin V, Vassart G, Froguel P, Refetoff S: Identification of a locus for nongoitrous congenital hypothyroidism on chromosome 15q25.3-26.1. Hum Genet 2005;118:348–355.
60 Lanzerath K, Bettendorf M, Haag C, Kneppo C, Schulze E, Grulich-Henn J: Screening for Pax8 mutations in patients with congenital hypothyroidism in South-West Germany. Horm Res 2006;66: 96–100.
61 Lapi P, Macchia PE, Chiovato L, Biffali E, Moschini L, Larizza D, Baserga M, Pinchera A, Fenzi G, Di Lauro R: Mutations in the gene encoding thyroid transcription factor-1 (TTF-1) are not a frequent cause of congenital hypothyroidism (CH) with thyroid dysgenesis. Thyroid 1997;7: 383–387.
62 Perna MG, Civitareale D, De Fillipis V, Sacco M, Cisternino C, Tassi V: Absence of mutations in the gene encoding thyroid transcription factor-1 (TTF-1) in patients with thyroid dysgenesis. Thyroid 1997;7:377–381.

63 Tonacchera M, Banco M, Lapi P, Di Cosmo C, Perri A, Montanelli L, Moschini L, Gatti G, Gandini D, Massei A, Agretti P, De Marco G, Vitti P, Chiovato L, Pinchera A: Genetic analysis of TTF-2 gene in children with congenital hypothyroidism and cleft palate, congenital hypothyroidism, or isolated cleft palate. Thyroid 2004;14:584–588.
64 Giurgea I, Bellanne-Chantelot C, Ribeiro M, Hubert L, Sempoux C, Robert JJ, Blankenstein O, Hussain K, Brunelle F, Nihoul-Fekete C, Rahier J, Jaubert F, de Lonlay P: Molecular mechanisms of neonatal hyperinsulinism. Horm Res 2006;66:289–296.

Guy Van Vliet, MD
Endocrinology Service and Research Center
Department of Pediatrics
Sainte-Justine Hospital
University of Montreal, 3175 Côte Ste-Catherine
Montreal, Que. H3T 1C5 (Canada)
Tel. +1 514 345 4735, Fax +1 514 345 4988
E-Mail guy.van-vliet@recherche-ste-justine.qc.ca

Van Vliet G, Polak M (eds): Thyroid Gland Development and Function.
Endocr Dev. Basel, Karger, 2007, vol 10, pp 43–61

Thyroid Imaging in Children

Catherine Garel[a], *Juliane Léger*[b]

Departments of [a]Paediatric Imaging and [b]Paediatric Endocrinology, Hôpital Robert Debré, Assistance Publique-Hôpitaux de Paris, Paris, France

Abstract

Ultrasonography (US) and radionuclide thyroid scanning are the imaging modalities of choice in the evaluation of the thyroid gland in children. The main normal US patterns of the thyroid gland are reviewed. In early infancy, thyroid imaging is usually performed in the context of congenital hypothyroidism (CH). The aetiological diagnosis of CH is usually based on the combined use of US and radionuclide thyroid scanning. A detailed description of thyroid abnormalities observed in CH is given. Thyroid dysgenesis includes athyreosis, 'empty' thyroid area with ectopic thyroid tissue and thyroid hypoplasia (global hypoplasia and thyroid hemiagenesis). Mild thyroid developmental anomalies (pyramidal lobe, thyroglossal duct cysts, thyroid hemiagenesis) may be observed in euthyroid patients or among first-degree relatives of a CH population with thyroid dysgenesis. A normally located gland (goitre, normal-sized thyroid gland or hypoplastic thyroid gland) may also be observed in patients with CH. The main patterns observed in chronic and acute thyroiditis, in hyperthyroidism and thyroid tumours are described. Moreover, fetuses and neonates born to mothers with Graves' disease are at risk of thyroid dysfunction and goitre due to hypothyroidism in relation to excessive maternal treatment or to maternal hyperthyroidism which may be observed with fetal US.

Ultrasonography (US) and radionuclide thyroid scanning are the imaging modalities of choice in the evaluation of the thyroid gland in children. A good knowledge of thyroid embryology is required in order to understand the different patterns observed in congenital hypothyroidism (CH) as well as in certain thyroid developmental anomalies (TDAs). Some maternal thyroid hormones, antibodies and antithyroid drugs cross the placenta, so that maternal thyroid disease may induce thyroid dysfunction in fetuses and neonates. In early infancy, thyroid imaging is usually performed in the context of CH, whereas later, during childhood, imaging is more focused on the evaluation of thyroid nodules or goitre in relation with hyperthyroidism (Graves' disease), hypothyroidism or euthyroidism (thyroiditis).

Normal Appearance of the Thyroid Gland on US

Technique

Patients are examined in the supine position with their neck well extended. Images are obtained with a high frequency probe (10–15 MHz) in transverse and longitudinal planes. A study of the thyroid gland by US must always include a systematic survey of the whole anterior cervical area from the foramen caecum (base of the tongue) to the normal anatomical position of the thyroid gland and even lower above the sternal manubrium.

The thyroid gland is easily identified at the lower part of the neck and appears as a homogeneous bilobed structure. Its echogenicity is greater than that of the neck muscles. Colloid follicles, appearing as small cystic areas (1–3 mm in diameter), are occasionally seen within the parenchyma. The follicles may contain echogenic foci representing the colloid. The lobes are bordered laterally by the common carotid artery and the internal jugular vein and medially by the trachea. Anterior to each lobe are the anterior muscles of the neck and posterior to the thyroid are the longus colli muscles. The oesophagus, with its echogenic mucosa, is visible medially, adjacent to the trachea [1, 2].

In the neonate, the thyroid lobes appear as ovoids. In the older child, their shape is more triangular [3] (fig. 1).

US volumetry is considered the most reliable method for determining thyroid volume. The volume of a thyroid lobe is usually calculated by using the formula: depth $\times$ length $\times$ width $\times$ $\pi/6$. The total thyroid volume is obtained by adding up the volumes of both lobes. The isthmus is not taken into account. Thyroid volume is best correlated with the body surface area [3]. Interobserver errors may occur to a rate of up to 30% in the assessment of thyroid volume by US, with the widest variation encountered in determining the length (craniocaudal diameter) of the thyroid gland. Intraobserver variation is smaller (about 8%) [4].

Normal values have been established in children [3, 5]. At all ages, the volume of the right lobe is slightly higher than that of the left lobe. In neonates, the volume ranges from 0.38 to 1.42 ml (mean volume = 0.84). From 7 days to 8 years, the volume of the gland increases slowly, with no significant difference between girls and boys. Above the age of 8, the volume increases more rapidly, approaching adult values at the end of puberty (mean value = 2.7 ml in prepubertal children and 11.6 ml in late pubertal patients over 17 years) [6]. Thyroid growth in adolescents with a normal iodine supply is mainly influenced by growth factors involved in somatic development, and is further modulated by

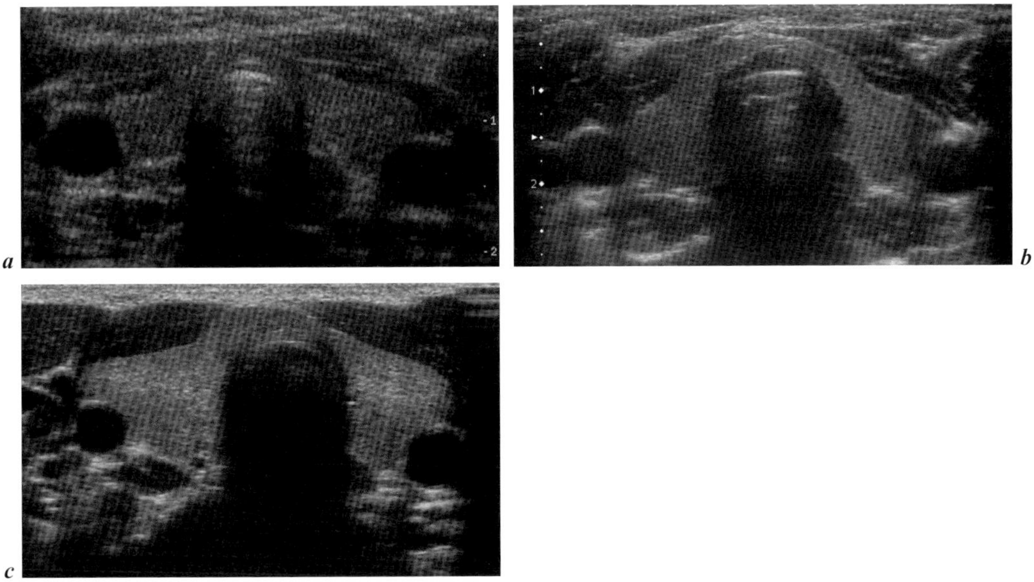

Fig. 1. Normal sonographic appearance of the thyroid gland in a neonate: width = 7 mm, depth = 5 mm, height = 20 mm (***a***); a 6-year-old boy: width = 10 mm, depth = 10 mm, height = 30 mm (***b***), and a 15-year-old boy: width = 16 mm, depth = 11.5 mm, height = 40 mm (***c***).

sex steroids. Specifically, the effect of oestrogens may account for the higher incidence of goitre in girls with mild iodine deficiency [7].

Modest differences in iodine intake may lead to marked differences in the incidence of goitre; when iodine supplementation is implemented, thyroid enlargement is readily reversible in children [8].

Developmental Abnormalities of the Thyroid Gland with or without Hypothyroidism

CH is a relatively common disorder, occurring in approximately 1 in 3,000 to 1 in 4,000 live births [9, 10]. Patients with CH are classified as having thyroid dysgenesis (85% of cases) or as having a normally located gland with or without goitre.

The aetiological diagnosis of CH is usually based on the combined use of US and radionuclide thyroid scanning. Both modalities are complementary tools. A detailed description of thyroid abnormalities is essential because it

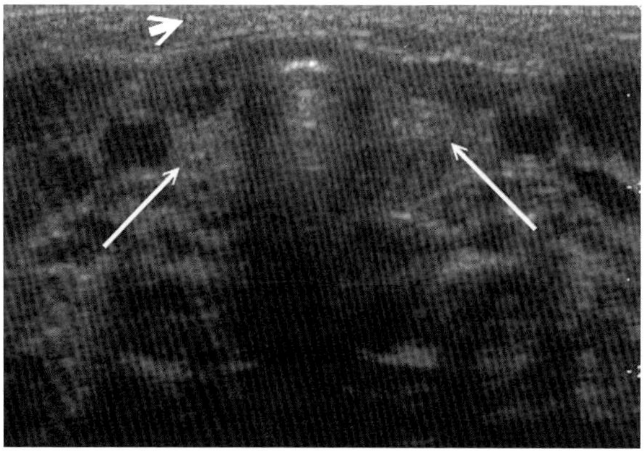

Fig. 2. 'Empty' thyroid area in a neonate with ectopic thyroid tissue. Hyperechoic structures (arrows; width = 7 mm; depth = 5 mm) are visible on both sides of the trachea. Their echogenicity is approximately the same as that of fat (arrowhead).

contributes to widening the knowledge of CH, and also because it plays an important role in determining disease severity and outcome.

Thyroid dysgenesis comprises a heterogeneous group of TDAs. It includes athyreosis (no thyroid tissue can be seen in normal or ectopic location), ectopic thyroid tissue without iodine uptake in the thyroid area, and thyroid hypoplasia (global hypoplasia and thyroid hemiagenesis).

US correctly establishes whether thyroid tissue is present in its normal location. Normal thyroid tissue is more echogenic than the muscles but less echogenic than fat. In the absence of thyroid lobes, small hyperechoic structures are located laterally on both sides of the trachea. Their appearance clearly differs from thyroid tissue because they are smaller than the normal thyroid lobes (about 5 mm deep and wide), do not increase in size during childhood and, more importantly, have a marked hyperechogenicity. The echogenicity of these structures is approximately the same as that of fat [6, 11–13] (fig. 2). Of course, the older the child, the easier the diagnosis with US, since the volume of these structures remains small and stable, but trained sonographers can quite easily establish this diagnosis even in neonates. It has been suggested that these structures could represent remnants of the ultimobranchial bodies, which are known to contain calcitonin-secreting cells [6].

In a recent series of 57 cases of children with CH due to thyroid dysgenesis diagnosed during the neonatal period, a prospective evaluation by US was performed at the age of 10.5 ± 4.5 years. Cysts were found in the 'empty' thyroid area in 68% of patients presenting with either ectopic thyroid tissue,

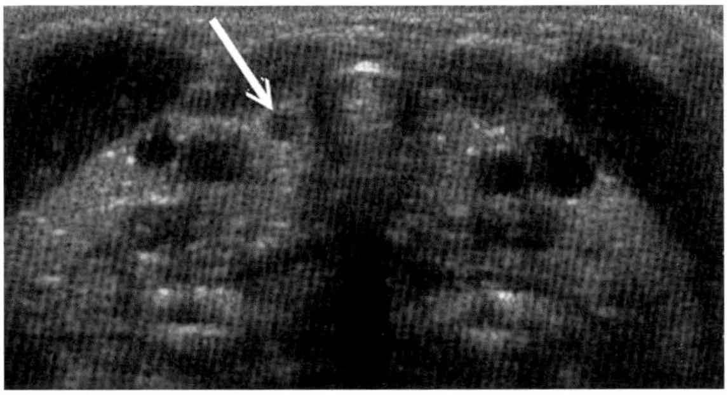

Fig. 3. 'Empty' thyroid area in a 1-month-old girl with athyreosis. One cyst (arrow) is visible within the 'empty' thyroid area.

athyreosis, or hemiagenesis. They were located close to the midline, were single or multiple, uni- or bilateral, vertically oval or round, with a mean size of 3.5 mm. They were unrelated to the age of the patients or the quality of treatment [12] (fig. 3). The exact nature of these cysts is unknown. They may result from the persistence of ultimobranchial bodies as a cystic structure. They might also be due to a persistence of the thyroglossal duct with cystic degeneration of the thyroid follicular cells that have completed their normal migration, even if the thyroid tissue remains incompletely descended or the thyroid gland has disappeared entirely [12].

Additional thymus tissue has also been observed within the 'empty' thyroid area in 4 out 42 patients with athyreosis or ectopic thyroid tissue. The precise histological nature of this tissue has not been established with certainty, but its echogenicity was exactly the same as that of the thymus: the structure was hypoechoic with multiple linear echoes and discrete echogenic foci. The thymus could be seen in its normal location in all the children. During embryonic life, the thymus glands migrate to their definitive location inferior and ventral to the developing thyroid and fuse to form a single, bilobed thymus structure. Occasionally during this descent, remnants of thymic tissue can be implanted along the cervical pathway [13].

Whether the molecular mechanisms governing development, morphogenesis, or migration of the embryonic thyroid and thymus share dependent factors remains to be explored.

In children with CH and thyroid dysgenesis, 'empty' thyroid area can also be associated with ectopic thyroid tissue, which is the most frequent cause of CH, accounting for approximately two thirds of cases. Ectopic or maldescended thyroid tissue represents an arrest in the usual descent of the thyroid tissue. It is not uncommon, and the presence of lingual thyroid tissue has been demonstrated in 10% of autopsy subjects who had a thyroid gland in the normal location [14].

A significant proportion of ectopic thyroid glands is also found incidentally in asymptomatic patients, thus suggesting that many are never diagnosed [15, 16]. These patients present with an 'empty' thyroid area, and ectopic tissue is observed at any site along the thryoglossal duct pathway. The presence of such ectopic thyroid tissue emphasizes the necessity to obtain imaging of the thyroid area in all patients with an anterior neck mass before removing it, in order to avoid hypothyroidism as a consequence of surgical excision [17].

The diagnosis of ectopic thyroid tissue can be established by radionuclide thyroid scanning (ectopic thyroid tissue shows a marked iodine uptake), US or magnetic resonance imaging. With US, it is mandatory to systematically survey the entire anterior cervical area, including the base of the tongue. The ectopic tissue demonstrates an echogenicity and vascularity similar to those of the normal thyroid gland. Any tissue of homogeneous appearance lying along the normal pathway of the thyroglossal duct is considered to be suggestive of ectopic thyroid tissue. The tissue must be assessed in size and echogenicity (compared with muscle) and its location must be defined in relation to the hyoid bone. In addition, the vascularity of this tissue must be evaluated with colour and/or power Doppler imaging. Using this method, ectopic thyroid tissue can be detected in about one quarter of children with CH and thyroid ectopia [18]. This detection rate is far lower than with radionuclide scanning but higher than previously reported in older US series, which have certainly led to a marked underestimation of the possibilities of US because of the poor quality of US units at that time [11, 19–22]. It is interesting to note that the detection of ectopic thyroid tissue by US (possibly related to a larger amount of thyroid tissue) seems to correspond to less severe forms of CH, as demonstrated by the higher epiphyseal knee surface and higher mean serum FT_4 and FT_3 levels [18]. Ectopic thyroid tissue is visible with the same rate of detection in neonates and older children. In two thirds of the patients, the ectopic thyroid tissue is located at the suprahyoid level. Moreover, US reveals different echogenicity and vascularity in patients with visible ectopic thyroid tissue, depending on thyroid function. In neonates, prior to treatment for CH, ectopic tissue appears hyperechoic and hypervascular in Doppler US, i.e. identical to that observed in the thyroid gland of normal neonates (fig. 4). Conversely, in older children who have been treated for several years, ectopic thyroid tissue is mainly hypoechoic with no vascularity (fig. 5). Reduction in blood flow may correspond to the normalization of TSH values after treatment, hence the reduction in stimulation of remnant thyroid tissue and the decreased activity. Variations in echogenicity could also be related to variations in activity [18]. The presence of a double ectopia is a very rare condition that has been reported by very few authors [18, 23–25].

Ectopic thyroid tissue can also be detected by magnetic resonance imaging. It appears as a mass of higher intensity than the surrounding tissue on both

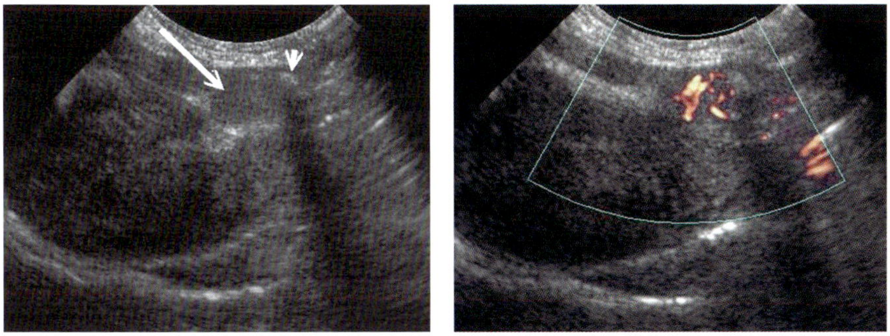

a b

Fig. 4. Ectopic thyroid tissue in a neonate. ***a*** The ectopic tissue (arrow) is hyperechoic and is located below the hyoid bone (arrowhead) which is well depicted on this midline sagittal slice. ***b*** Hypervascularization of the thyroid ectopic tissue is observed with colour Doppler US.

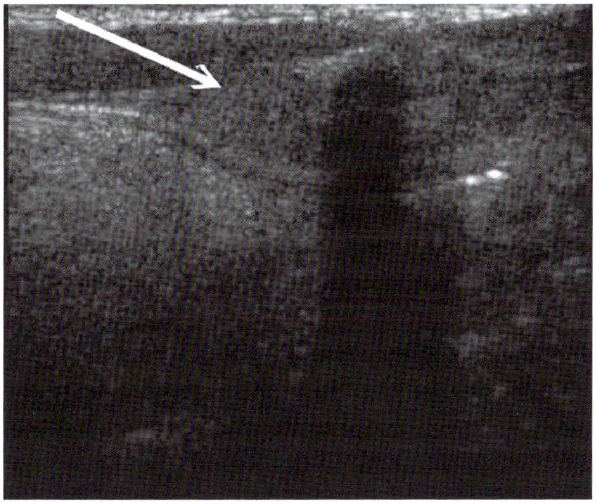

Fig. 5. Thyroid ectopic tissue in a 14-year-old girl treated for CH. Midline sagittal slice at the level of the base of the tongue. The thyroid ectopic tissue is hypoechoic (arrow).

T1-weighted and T2-weighted images. Few cases have been reported in adults and children [22, 26–28]. The usefulness of this imaging modality in the aetiological diagnosis of CH has not been studied.

Mild TDAs, including ectopic thyroid tissue as mentioned above, can be observed in euthyroid patients or among first-degree relatives of a CH population with thyroid dysgenesis [29]. US is superior to radionuclide thyroid scanning in the assessment of mild TDAs [13].

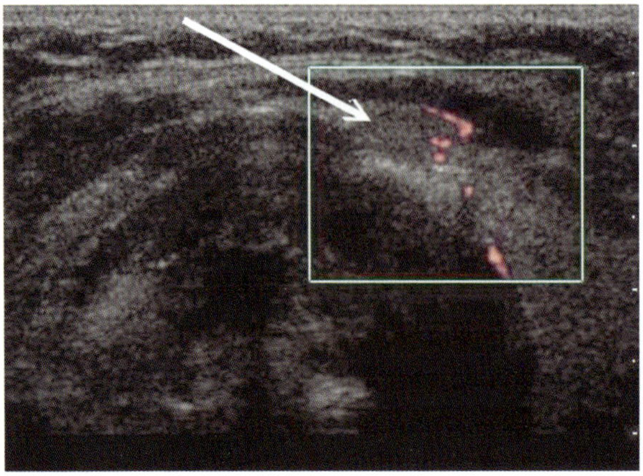

Fig. 6. Pyramidal lobe detected with US in the mother of a child treated for CH with athyreosis. This additional thyroid tissue (arrow) is attached to the left thyroid gland and has the same echogenicity as normal thyroid tissue.

A pyramidal lobe represents the persistence of the caudal portion of the thyroglossal duct. It appears as additional thyroid tissue, usually lying in the midline and attached to the thyroid gland, but it can also arise from either lobe (more commonly from the left lobe). It is isoechoic to the thyroid gland and shows iodine uptake at radionuclide thyroid scanning [29] (fig. 6).

Thyroglossal duct cysts are located at any site along the pathway of the thyroglossal duct. They result from cell residues remaining along this pathway. Familial cases of thyroglossal duct cysts have been reported [30–32]. On US, these cysts are anechoic (unless they are infected) and are located on the midline or in a parasagittal location (fig. 7).

Thyroid hemiagenesis is a disorder thought to result either from a failure of the thyroid anlage to become bilobed and to spread out on both sides, or from involution of one side of the bilobed structure. It is believed to be the rarest of all TDAs, its prevalence during routine thyroid US evaluation being estimated between 0.05 and 0.2% [33–35]. The missing lobe is usually the left one (three quarters of cases) and the isthmus is absent in 60% of cases (fig. 8). Hypoplasia of one lobe can also be observed (fig. 9). An enlargement of the existing lobe has been described in some series [34], but no major compensatory enlargement was found in other studies [33, 35]. US has been found to be clearly superior over radionuclide thyroid scanning for the diagnosis of thyroid hemiagenesis [33]. Thyroid function is within the normal range in the majority of patients but hypothyroidism can also be observed in those patients as mentioned above [35, 36]. Familial forms of thyroid hemiagenesis have been reported [35, 37].

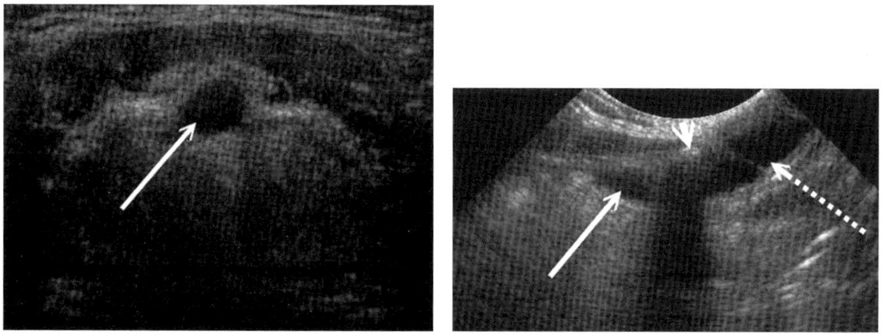

a b

Fig. 7. Intralingual thyroglossal duct cyst (arrow) in a 7-year-old euthyroid boy. ***a*** Coronal slice at the level of the base of the tongue. ***b*** Midline sagittal slice at the same level. The hyoid bone (arrowhead) is hyperechoic and the subhyoid component of the cyst (dotted arrow) is visible.

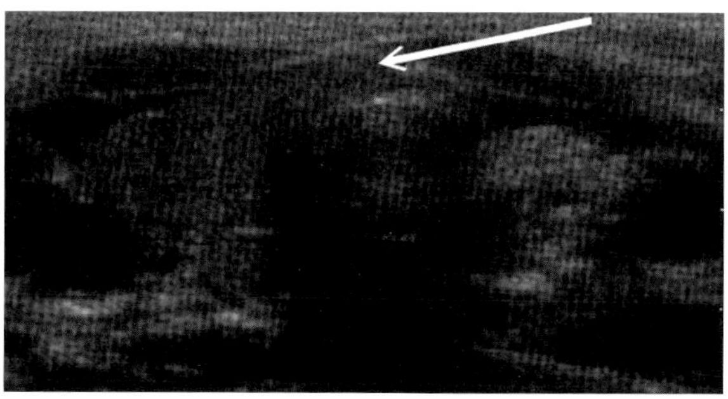

Fig. 8. Thyroid hemiagenesis. The left lobe is absent but the isthmus (arrow) is present.

The pathogenesis of TDAs is unknown and TDAs have usually been considered sporadic. Associations among these anomalies have been reported: hemiagenesis accompanying an ectopic gland [38], or a thyroglossal duct cyst [39], or an ectopic thyroid with thyroglossal duct cyst [40]. The existence of such associations, the presence of familial forms and the report of TDAs among first-degree asymptomatic relatives of a CH population with thyroid dysgenesis [29] argue in favour of genetically determined mechanisms. When thyroid dysgenesis is diagnosed in a child presenting with CH, US in first-degree relatives may reveal asymptomatic TDAs.

A normally located gland is observed in 15% of patients with CH and represents a heterogeneous group. The thyroid gland may be hypertrophied (goitre), have a normal size, or be globally hypoplastic. In addition, in this group of patients, hypothyroidism may be permanent or transient. In the group

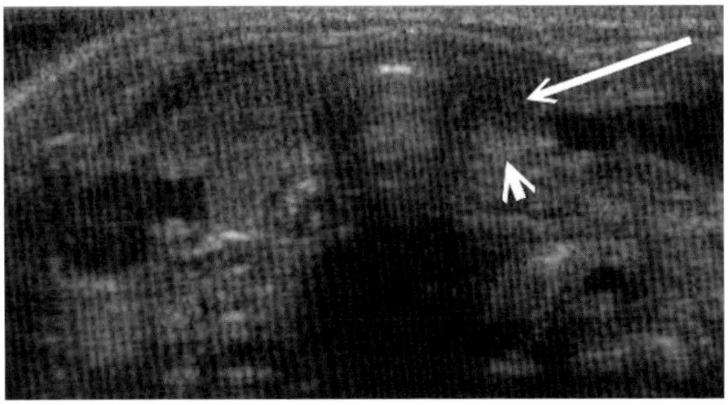

Fig. 9. Hypoplasia of the left thyroid lobe in a neonate. The left lobe (arrow) is much smaller than the right one. Behind the left lobe, the same hyperechogenicity as in 'empty' thyroid area can be seen (arrowhead).

with transient hypothyroidism, the most common causes are iodine deficiency (the prevalence of which varies widely in the world), iodine overload, particularly in premature newborns, or transplacental passage of antithyroid antibodies or antithyroid drugs. In a recent study of 79 patients with CH and a normally located thyroid gland, transient hypothyroidism was observed in 38% of cases. Among permanent CH cases, 55% had goitre, 29% had a normal-sized and shaped thyroid gland and 16% showed a hypoplastic gland (either global or partial hypoplasia or asymmetry of the two lobes or hemiagenesis). Patients with a normal-sized and shaped thyroid gland had a significantly less severe form of hypothyroidism than those with goitre or a hypoplastic thyroid gland [36].

Goitre is usually associated with inborn abnormalities of thyroid hormone synthesis, iodine organification defects and defects of thyroglobulin synthesis being the most frequent causes. This type of disease, which has an autosomal recessive mode of inheritance, is assessed by means of a thyroid radionuclide scanning with perchlorate discharge test. Pendred disease has also been reported in association with CH and goitre, but in 25% of cases no aetiology can be determined [36]. An absence of iodine uptake can also be observed in patients with goitre related to a defect of the sodium/iodine symporter [36]. Nodules may develop within goitre related to enzyme deficiencies, and make it necessary to follow up these children with US [41].

Permanent CH with a normal-sized thyroid gland can be observed in pseudohypoparathyroidism and Down's syndrome [36].

In several patients with permanent CH and normal or hypoplastic thyroid gland, inactivating mutations in the TSH receptor gene have been found [36, 42–44]. In these patients with severe TSH resistance, no iodine uptake is observed with radionuclide scanning despite the presence of a thyroid gland;

therefore, this entity can be misdiagnosed as a true athyreosis, which emphasizes the necessity of combining radionuclide scanning and US in the evaluation of CH patients [45].

Regarding the role of both US and radionuclide scanning in the aetiological diagnosis of CH, it can be said that:

- US is useful in identifying a normally located thyroid gland and in delineating its anatomy, but it is less accurate than radionuclide thyroid scanning in the diagnosis of ectopic glands.
- Radioiodine thyroid scanning is still required in the clinical evaluation of these patients, especially of patients where, with US, the thyroid gland cannot be seen in its normal location, and in patients with goitre in whom a positive perchlorate discharge test makes it possible to diagnose an organification defect [13].`

Chronic Autoimmune Thyroiditis

Chronic autoimmune thyroiditis (or Hashimoto's thyroiditis) is defined as the association of circulating thyroid antibodies with a morphological abnormality of the thyroid gland or hypothyroidism. Its prevalence is higher in girls than in boys and a family history of thyroid disease is present in about one quarter of cases. Other autoimmune diseases are found in one fifth of the children. At initial presentation, goitres are observed in 80% of cases. Imaging by US or radionuclide scanning is usually not required to make a diagnosis of Hashimoto's thyroiditis. If US is performed, it will show an enlarged, hypoechoic gland (compared to muscles) with a coarse heterogeneous echogenicity. These findings are more commonly observed in children with hypothyroidism. Nodules are present in one fifth of the children [46–49] (fig. 10). Radionuclide scanning may reveal inhomogeneous uptake, but is generally normal [48].

Acute Thyroiditis

Acute thyroiditis is a rare condition in children. It may be related to a congenital fistula between the pyriform sinus and the ipsilateral lobe of the thyroid gland or the perithyroidal space. Such a fistula is considered a remnant of the fourth pharyngeal pouch [50]. Congenital fistulas are more common on the left side than on the right. On US, the thyroid gland is enlarged and heterogeneous with multiple hypoechoic and complex masses. Abscess formation and involvement of the perithyroid area can also be observed [1, 51, 52] (fig. 11). The diagnosis can be established by endoscopy showing the fistula at the bottom of the pyriform sinus.

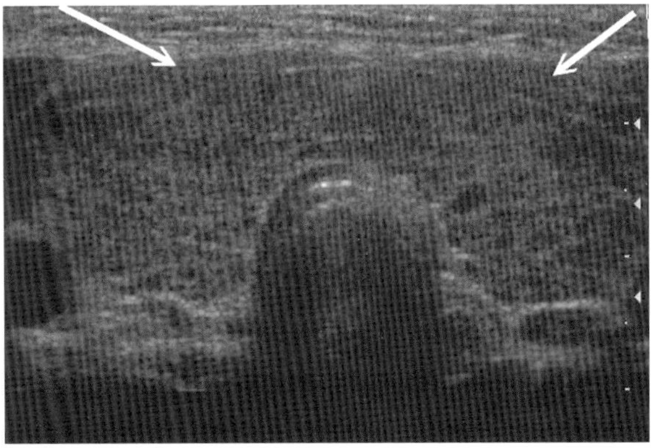

Fig. 10. Hashimoto's thyroiditis associated with diabetes mellitus in an 11-year-old girl. The thyroid gland is enlarged and hypoechoic compared with the surrounding muscles (arrows). It presents a coarse heterogeneous echogenicity.

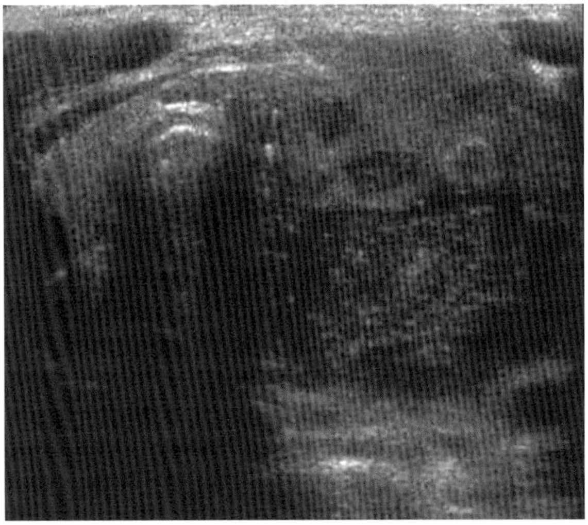

Fig. 11. Acute thyroiditis of the left thyroid lobe in relation with a fistula between this lobe and the ipsilateral pyriform sinus. The left thyroid lobe is enlarged, hypoechoic and heterogeneous. The borders of the lobe are ill defined because of the involvement of the perithyroid area.

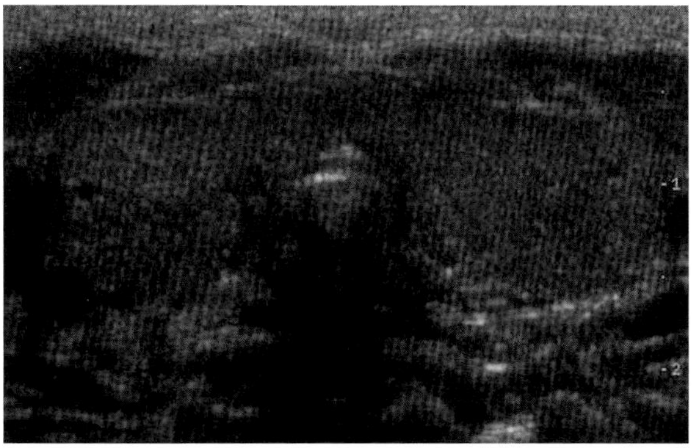

Fig. 12. Homogeneous goitre in a neonate born to a mother with Graves' disease.

Hyperthyroidism

Fetuses and neonates born to mothers with Graves' disease are at risk of thyroid dysfunction if the mother tests positive for TSH receptor antibodies and/or receives antithyroid drugs during pregnancy. The fetal thyroid gland starts secreting thyroid hormones at about 12 weeks of development, and fetal TSH receptors become responsive to TSH around 20 weeks. Before 12 weeks, the mother is the only source of thyroid hormones for the fetus. Maternal T_4 crosses the placenta to some extent while TSH does not cross the placenta at all. The maternal thyroid gland has to adjust the output of thyroid hormones to the state of pregnancy and maintain this equilibrium until term [53, 54].

Both TSH receptor antibodies and antithyroid drugs cross the placenta. One quarter of the fetuses born to high-risk mothers present goitre on fetal US (fig. 12). In these fetuses, the main issue is to determine whether goitre is due to hypothyroidism in relation to excessive maternal treatment or related to hyperthyroidism, the fetal thyroid gland being stimulated by TSH receptor antibodies due to maternal Graves' disease. Doppler examination of the fetal thyroid gland proves useful since, when hypervascularization is confined to the periphery of the gland, the pattern is suggestive of hypothyroidism [53].

Graves' disease is an autoimmune disease caused by thyroid-stimulated immunoglobulin binding to the TSH receptor. It is more frequent in girls than in boys and peaks in adolescence. Sonographically, the thyroid gland is diffusely enlarged, often with lobulated contours. The gland may show a normal echogenicity or may be hypoechoic as in thyroiditis (fig. 13). Increased diffuse parenchymal hypervascularity (named 'thyroid inferno') is observed [1, 55]. High-graded hypervascularization is not observed to such an extent in patients

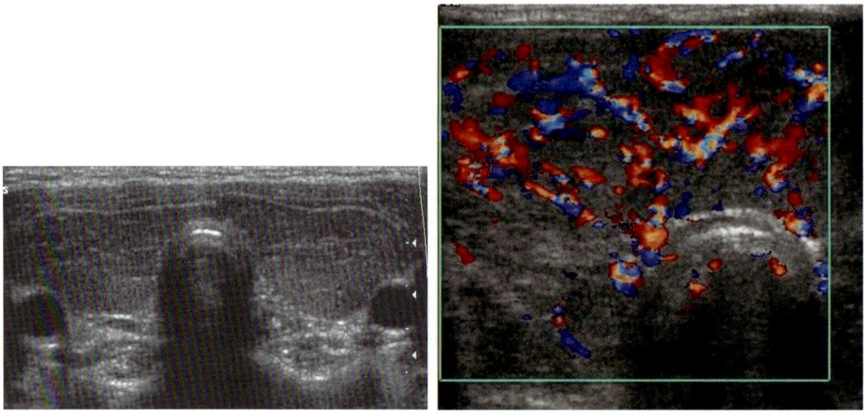

a b

Fig. 13. Graves' disease in a 5-year-old boy. ***a*** The thyroid gland is enlarged and diffusely hypoechoic. ***b*** Marked hypervascularization of the thyroid gland is demonstrated with colour Doppler US.

with chronic autoimmune thyroiditis [56]. The importance of the goitre is variable and may be small, moderate or large [57]. In exceptional cases, the volume of the thyroid is normal.

Thyroid Nodules

Thyroid nodules are rare in children as compared to adults. Benign follicular adenoma is the most common cause of thyroid nodules in children. Most nodules are hypoechoic compared with the rest of the gland [1], but iso- and hyperechoic nodules may also be observed. A peripheral hypoechoic halo is commonly seen around adenomas: it is caused by the fibrous capsule and blood vessels [51]. Some toxic adenomas may be responsible for associated hyperthyroidism. A circumscribed hypervascularization is observed with colour Doppler US [55] (fig. 14). Fine-needle biopsy has a 90–95% accuracy of diagnosis in children [51].

Thyroid Tumours

Cervical teratomas are usually large masses easily recognizable at birth; they can also be detected with US in fetuses (fig. 15). They may contain fat, teeth and soft-tissue elements and have a non-homogeneous echogenicity. They develop within or adjacent to the thyroid gland [1, 58].

Carcinoma of the thyroid is uncommon in children. On US, the thyroid gland shows a hypoechoic solid mass, sometimes with coarse calcifications

14

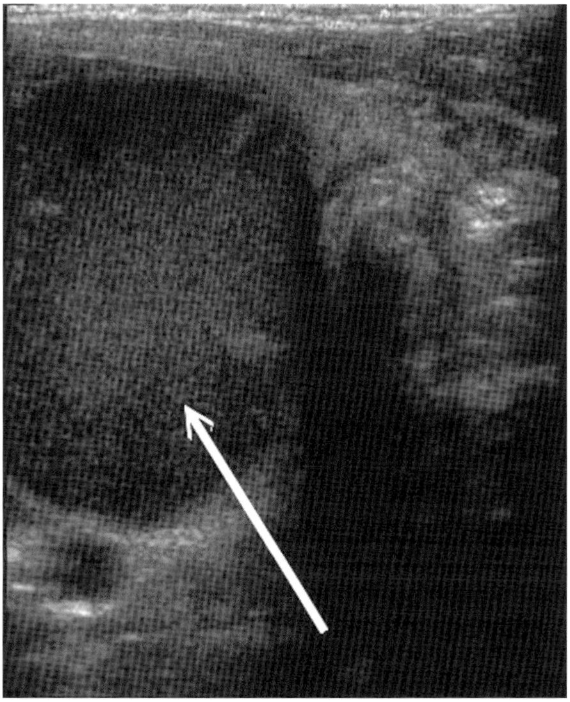

15

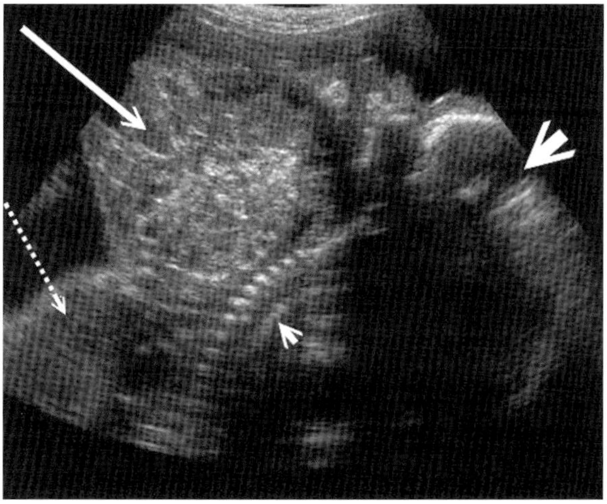

Fig. 14. Huge thyroid nodule developed in the right lobe. The nodule (arrow) is hypoechoic.

Fig. 15. Cervical teratoma developed in the right thyroid lobe of a 27-week-old fetus. Midline sagittal slice. This huge heterogeneous mass was responsible for polyhydramnios. The teratoma (arrow) is visible above the thorax (dotted arrow), before the cervical spine (small arrowhead). The fetal head is indicated by the large arrowhead.

[51]. Papillary carcinoma [see chapter by Vasko et al., pp. 140–172] is more common than medullary carcinoma. In paediatric patients, the latter is generally familial and observed in children of parents with multiple endocrine neoplasia type 2 [see chapter by Szinnai et al., pp. 173–187].

Conclusion

US is a non-invasive imaging tool, and is superior to radionuclide scanning in the assessment of normally located thyroid glands, as it provides a good evaluation of their anatomy and echogenicity. US is less accurate than radionuclide scanning in the diagnosis of ectopic thyroid tissue but gives a more detailed description (location, echogenicity, vascularization) of such tissue.

The ultrasound findings in some thyroid diseases (chronic autoimmune thyroiditis, Graves' disease, thyroid nodules) are similar in children and adults. Conversely, some patterns (mainly those related to CH) are discovered mainly during the neonatal period. The pathogenesis of a vast majority of these diseases remains unknown, even though a genetic implication has been proved in some cases. Screening for TDAs among first-degree relatives of CH patients might be useful to evaluate the familial component of the disease. A detailed description of thyroid abnormalities in CH is essential, in order to improve the knowledge of the disease, and also to predict its permanence and severity.

References

1 Siegel M: Face and neck; in Siegel M (ed): Pediatric Sonography. Philadelphia, Lippincott Williams & Wilkins, 2002, pp 123–166.

2 Butch RJ, Simeone JF, Mueller PR: Thyroid and parathyroid ultrasonography. Radiol Clin North Am 1985;23:57–71.

3 Chanoine JP, Toppet V, Lagasse R, Spehl M, Delange F: Determination of thyroid volume by ultrasound from the neonatal period to late adolescence. Eur J Pediatr 1991;150:395–399.

4 Ozgen A, Erol C, Kaya A, Ozmen MN, Akata D, Akhan O: Interobserver and intraobserver variations in sonographic measurement of thyroid volume in children. Eur J Endocrinol 1999;140:328–331.

5 Zimmermann MB, Hess SY, Molinari L, De Benoist B, Delange F, Braverman LE, Fujieda K, Ito Y, Jooste PL, Moosa K, Pearce EN, Pretell EA, Shishiba Y: New reference values for thyroid volume by ultrasound in iodine-sufficient schoolchildren: A World Health Organization/Nutrition for Health and Development Iodine Deficiency Study Group Report. Am J Clin Nutr 2004;79:231–237.

6 Chanoine JP, Toppet V, Body JJ, Van Vliet G, Lagasse R, Bourdoux P, Spehl M, Delange F: Contribution of thyroid ultrasound and serum calcitonin to the diagnosis of congenital hypothyroidism. J Endocrinol Invest 1990;13:103–109.

7 Fleury Y, Van Melle G, Woringer V, Gaillard RC, Portmann L: Sex-dependent variations and timing of thyroid growth during puberty. J Clin Endocrinol Metab 2001;86:750–754.

8 Knudsen N, Bulow I, Jorgensen T, Laurberg P, Ovesen L, Perrild H: Goitre prevalence and thyroid abnormalities at ultrasonography: a comparative epidemiological study in two regions with slightly different iodine status. Clin Endocrinol (Oxf) 2000;53:479–485.
9 Delange F: Neonatal screening for congenital hypothyroidism: results and perspectives. Horm Res 1997;48:51–61.
10 Kaplan EL, Shukla M, Hara H, Ito K: Developmental abnormalities of the thyroid; in De Groot LJ (ed): Endocrinology. Philadelphia, Saunders, 1994, pp 893–899.
11 Poyhonen L, Lenko HL: Ultrasonography in congenital hypothyreosis. Acta Paediatr Scand 1984;73:523–526.
12 Marinovic D, Garel C, Czernichow P, Leger J: Additional phenotypic abnormalities with presence of cysts within the 'empty' thyroid area in patients with congenital hypothyroidism with thyroid dysgenesis. J Clin Endocrinol Metab 2003;88:1212–1216.
13 Bubuteishvili L, Garel C, Czernichow P, Leger J: Thyroid abnormalities by ultrasonography in neonates with congenital hypothyroidism. J Pediatr 2003;143:759–764.
14 Sauk JJ Jr: Ectopic lingual thyroid. J Pathol 1970;102:239–243.
15 Hung W, Randolph JG, Sabatini D, Winship T: Lingual and sublingual thyroid glands in euthyroid children. Pediatrics 1966;38:647–651.
16 Gillis D, Brnjac L, Perlman K, Sochett EB, Daneman D: Frequency and characteristics of lingual thyroid not detected by screening. J Pediatr Endocrinol Metab 1998;11:229–233.
17 Holland AJ, Sparnon AL, LeQuesne GW: Thyroglossal duct cyst or ectopic thyroid gland? J Paediatr Child Health 1997;33:346–348.
18 Marinovic D, Garel C, Czernichow P, Leger J: Ultrasonographic assessment of the ectopic thyroid tissue in children with congenital hypothyroidism. Pediatr Radiol 2004;34:109–113.
19 Muir A, Daneman D, Daneman A, Ehrlich R: Thyroid scanning, ultrasound, and serum thyroglobulin in determining the origin of congenital hypothyroidism. Am J Dis Child 1988;142: 214–216.
20 De Bruyn R, Ng WK, Taylor J, Campbell F, Mitton SG, Dicks-Mireaux C, Grant DB: Neonatal hypothyroidism: comparison of radioisotope and ultrasound imaging in 54 cases. Acta Paediatr Scand 1990;79:1194–1198.
21 Ueda D, Mitamura R, Suzuki N, Yano K, Okuno A: Sonographic imaging of the thyroid gland in congenital hypothyroidism. Pediatr Radiol 1992;22:102–105.
22 Takashima S, Nomura N, Tanaka H, Itoh Y, Miki K, Harada T: Congenital hypothyroidism: assessment with ultrasound. AJNR Am J Neuroradiol 1995;16:1117–1123.
23 Hod N, Mindlin L, Cohenpour M, Horne T: Double ectopic thyroid. Pediatr Radiol 2002;32: 859–861.
24 Kumar R, Khullar S, Gupta R, Marwah A, Drm MA: Dual thyroid ectopy: case report and review of the literature. Clin Nucl Med 2000;25:253–254.
25 Hazarika P, Siddiqui SA, Pujary K, Shah P, Nayak DR, Balakrishnan R: Dual ectopic thyroid: a report of two cases. J Laryngol Otol 1998;112:393–395.
26 Ohnishi H, Sato H, Noda H, Inomata H, Sasaki N: Color Doppler ultrasonography: diagnosis of ectopic thyroid gland in patients with congenital hypothyroidism caused by thyroid dysgenesis. J Clin Endocrinol Metab 2003;88:5145–5149.
27 Johnson JC, Coleman LL: Magnetic resonance imaging of a lingual thyroid gland. Pediatr Radiol 1989;19:461–462.
28 Spinas GA, Staub JJ, Wey W, Nidecker A: Magnetic resonance imaging for the assessment of lingual thyroid. J Endocrinol Invest 1989;12:429–431.
29 Leger J, Marinovic D, Garel C, Bonaiti-Pellie C, Polak M, Czernichow P: Thyroid developmental anomalies in first degree relatives of children with congenital hypothyroidism. J Clin Endocrinol Metab 2002;87:575–580.
30 Ashworth J: Three generations of thyroglossal duct remnant in one family. J Fam Pract 1979;8:524–525.
31 Millikan JS, Murr P, Moore EE, Moore GE: A familial pattern of thyroglossal duct cysts. JAMA 1980;244:1714.
32 Issa MM, deVries P: Familial occurrence of thyroglossal duct cyst. J Pediatr Surg 1991;26:30–31.
33 Shabana W, Delange F, Freson M, Osteaux M, De Schepper J: Prevalence of thyroid hemiagenesis: ultrasound screening in normal children. Eur J Pediatr 2000;159:456–458.

34 Majorana R, Carta A, Floriddia G, Leonardi D, Buscema M, Sava L, Calaciura F, Vigneri R: Thyroid hemiagenesis: prevalence in normal children and effect on thyroid function. J Clin Endocrinol Metab 2003;88:1534–1536.

35 Castanet M, Leenhardt L, Leger J, Simon-Carre A, Lyonnet S, Pelet A, Czernichow P, Polak M: Thyroid hemiagenesis is a rare variant of thyroid dysgenesis with a familial component but without Pax8 mutations in a cohort of 22 cases. Pediatr Res 2005;57:908–913.

36 Gaudino R, Garel C, Czernichow P, Leger J: Proportion of various types of thyroid disorders among newborns with congenital hypothyroidism and normally located gland: a regional cohort study. Clin Endocrinol (Oxf) 2005;62:444–448.

37 Rajmil HO, Rodriguez-Espinosa J, Soldevila J, Ordonez-Llanos J: Thyroid hemiagenesis in two sisters. J Endocrinol Invest 1984;7:393–394.

38 Hsu CY, Wang SJ: Thyroid hemiagenesis accompanying an ectopic sublingual thyroid. Clin Nucl Med 1994;19:546.

39 Tsang SK, Maher J: Thyroid hemiagenesis accompanying a thyroglossal duct cyst: a case report. Clin Nucl Med 1998;23:229–232.

40 Wang CY, Chang TC: Preoperative thyroid ultrasonography and fine-needle aspiration cytology in ectopic thyroid. Am Surg 1995;61:1029–1031.

41 Niedziela M: Pathogenesis, diagnosis and management of thyroid nodules in children. Endocr Relat Cancer 2006;13:427–453.

42 Sunthornthepvarakui T, Gottschalk ME, Hayashi Y, Refetoff S: Brief report: resistance to thyrotropin caused by mutations in the thyrotropin-receptor gene. N Engl J Med 1995;332:155–160.

43 Biebermann H, Schoneberg T, Krude H, Schultz G, Gudermann T, Gruters A: Mutations of the human thyrotropin receptor gene causing thyroid hypoplasia and persistent congenital hypothyroidism. J Clin Endocrinol Metab 1997;82:3471–3480.

44 Alberti L, Proverbio MC, Costagliola S, Romoli R, Boldrighini B, Vigone MC, Weber G, Chiumello G, Beck-Peccoz P, Persani L: Germline mutations of TSH receptor gene as cause of nonautoimmune subclinical hypothyroidism. J Clin Endocrinol Metab 2002;87:2549–2555.

45 Castanet M, Polak M, Bonaiti-Pellie C, Lyonnet S, Czernichow P, Leger J: Nineteen years of national screening for congenital hypothyroidism: familial cases with thyroid dysgenesis suggest the involvement of genetic factors. J Clin Endocrinol Metab 2001;86:2009–2014.

46 Marinovic D, Leger J, Garel C, Czernichow P: Chronic autoimmune thyroiditis in the child. Arch Pediatr 2000;7:1284–1292.

47 Marcocci C, Vitti P, Cetani F, Catalano F, Concetti R, Pinchera A: Thyroid ultrasonography helps to identify patients with diffuse lymphocytic thyroiditis who are prone to develop hypothyroidism. J Clin Endocrinol Metab 1991;72:209–213.

48 Alos N, Huot C, Lambert R, Van Vliet G: Thyroid scintigraphy in children and adolescents with Hashimoto disease. J Pediatr 1995;127:951–953.

49 Ivarsson SA, Ericsson UB, Fredriksson B, Persson PH: Ultrasonic imaging in the differential diagnosis of diffuse thyroid disorders in children. Am J Dis Child 1989;143:1369–1372.

50 Benson MT, Dalen K, Mancuso AA, Kerr HH, Cacciarelli AA, Mafee MF: Congenital anomalies of the branchial apparatus: embryology and pathologic anatomy. Radiographics 1992;12: 943–960.

51 Babcock DS: Thyroid disease in the pediatric patient: emphasizing imaging with sonography. Pediatr Radiol 2006;36:299–308, quiz 372–393.

52 Lucaya J, Berdon WE, Enriquez G, Regas J, Carreno JC: Congenital pyriform sinus fistula: a cause of acute left-sided suppurative thyroiditis and neck abscess in children. Pediatr Radiol 1992;21:27–29.

53 Luton D, Le Gac I, Vuillard E, Castanet M, Guibourdenche J, Noel M, Toubert ME, Leger J, Boissinot C, Schlageter MH, Garel C, Tebeka B, Oury JF, Czernichow P, Polak M: Management of Graves' disease during pregnancy: the key role of fetal thyroid gland monitoring. J Clin Endocrinol Metab 2005;90:6093–6098.

54 Serreau R, Polak M, Leger J, Vuillard E, Thurninger O, Chemouny S, Heid M, Guibourdenche J, Jacqz-Aigrain E, Oury JF, Luton D: Fetal thyroid goiter after massive iodine exposure. Prenat Diagn 2004;24:751–753.

55 Bogazzi F, Bartalena L, Brogioni S, Burelli A, Manetti L, Tanda ML, Gasperi M, Martino E: Thyroid vascularity and blood flow are not dependent on serum thyroid hormone levels: studies in vivo by color flow Doppler sonography. Eur J Endocrinol 1999;140:452–456.
56 Schulz SL, Seeberger U, Hengstmann JH: Color Doppler sonography in hypothyroidism. Eur J Ultrasound 2003;16:183–189.
57 Glaser NS, Styne DM: Predictors of early remission of hyperthyroidism in children. J Clin Endocrinol Metab 1997;82:1719–1726.
58 Vazquez E, Enriquez G, Castellote A, Lucaya J, Creixell S, Aso C, Regas J: US, CT, and MR imaging of neck lesions in children. Radiographics 1995;15:105–122.

Catherine Garel, MD
Service d'Imagerie Pédiatrique, Hôpital Robert Debré
48 boulevard Sérurier
FR–75019 Paris (France)
Tel. +33 140 032 247, Fax +33 140 032 245, E-Mail catherine.garel@rdb.ap-hop-paris.fr

Van Vliet G, Polak M (eds): Thyroid Gland Development and Function.
Endocr Dev. Basel, Karger, 2007, vol 10, pp 62–85

Clinical and Biological Consequences of Iodine Deficiency during Pregnancy

Daniel Glinoer

Division of Endocrinology, Department of Internal Medicine, Thyroid Investigation Clinic, University Hospital Saint Pierre, Brussels, Belgium

Abstract

The main change in thyroid function associated with the pregnant state is the requirement of an increased production of thyroid hormone that depends directly upon the adequate availability of dietary iodine and integrity of the glandular machinery. In healthy pregnant women, physiological adaptation takes place when the iodine intake is adequate, while this is replaced by pathological alterations when there is a deficient iodine intake. Pregnancy acts typically, therefore, as a revelator of underlying iodine restriction. Iodine deficiency has important repercussions for both the mother and the fetus, leading to hypothyroxinemia, sustained glandular stimulation and finally goitrogenesis. Furthermore, because severe iodine deficiency may be associated with an impairment in the psychoneurointellectual outcome in the progeny, because both mother and offspring are exposed to iodine deficiency during gestation (and the postnatal period), and because iodine deficiency is still prevalent today in several large regions of the world, iodine supplements should be given systematically to pregnant and breastfeeding mothers. Particular attention is required to ensure that pregnant women receive an adequate iodine supply, in order to reach the ideal recommended nutrient intake of 250 μg iodine/day.

Introduction

Iodine deficiency (ID) is a major threat to the health and development of populations worldwide, with preschool children and pregnant women representing the target groups with the highest risks. When requirements for iodine are not met, thyroid hormone (TH) synthesis is impaired, resulting in a series of functional and developmental abnormalities collectively referred to as ID disorders. Depending upon the degree of severity of ID in a given population, clinical conditions associated therewith include goiter (referred to as 'endemic goiter'), miscarriage and still birth, hypothyroidism, impaired growth, as well as mental and neurological disorders resulting form irreversible brain damage (referred to as 'endemic cretinism'). Although cretinism is the most dramatic expression of severe ID, more subtle degrees of mental impairment with poor learning abilities in schoolchildren as well as reduced intellectual performance and impaired working capability are also of considerable significance [1].

For the thyroid gland, the pregnant state represents a prolonged condition associated with alterations in the regulation of thyroid function and economy, due to separate physiological events that take place at different time points during gestation, and constitute a challenge for the maternal thyroid, because together these events exert stimulatory effects on the glandular machinery [2]. Until the late 1980s, it was commonly accepted that the thyroid gland was capable of adapting physiologically to the pregnant state without much happening in healthy euthyroid women. The main 'apparent' changes in thyroid function tests – that had been identified since the early 1960s – were an increase in serum total T_4 and T_3 concentrations following the marked increase in serum thyroxine-binding globulin (TBG) concentrations, itself resulting from sustained estrogen stimulation [3]. In these early days, it was also considered that the production of TH was not modified during pregnancy [4]. Nowadays however, this view has been completely modified, as several important metabolic changes are known to take place during pregnancy.

Regulation of Thyroid Function during Pregnancy

Beginning already in early gestation, reaching a plateau at midgestation that is maintained thereafter until term, there is a 2- to 3-fold increase in serum TBG concentrations under the influence of a sustained rise in the concentration of estrogen [5]. Also starting in early gestation, there is an increase in renal blood flow and glomerular filtration which leads to an increased iodide clearance from plasma, and thus to an obligatory loss of iodine. Occurring transiently near the end of the first trimester, there is direct stimulation of the maternal thyroid gland by an increase in the concentration of human chorionic gonadotropin that may lead temporarily to a slight increase in the concentration of free T_4 [6, 7]. Finally, significant changes occur in the peripheral metabolism of maternal TH during the second half of gestation, mainly under the influence of placental type 3 iodothyronine deiodinase [8, 9]. Together, these events represent a profound metabolic change, associated with the progression of gestation already during its first half, which constitutes a transition from the preconception thyroidal steady state to the pregnancy steady state.

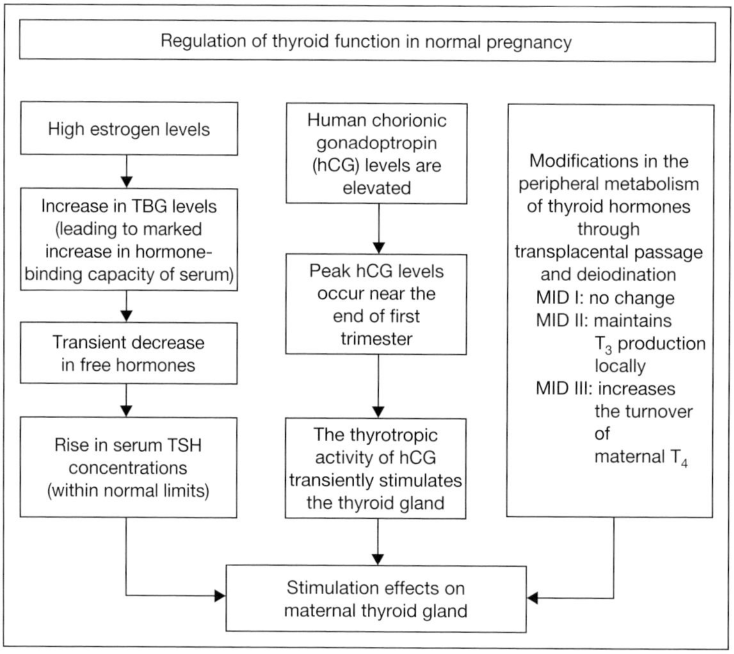

Fig. 1. The scheme shows the three series of separate events which together concur in exerting stimulatory effects on thyroid function during a normal pregnancy. The first event is linked to the progressive rise in TBG concentrations during the first trimester; the second event takes place transiently near the end of the first trimester and is related to the thyrotropic action of peak hCG concentrations; the third event takes place mainly during the second half of gestation and is related to pregnancy-specific modifications in peripheral metabolism of maternal TH, mainly at placental level. MID = Monoiodothyronine deiodinase (type I, II, III).

In order to be met, these metabolic changes require an increase in hormone production by the maternal thyroid gland (fig. 1). Once a new equilibrium has been reached, the increased demands of TH are sustained until term. For healthy pregnant women with a sufficient iodine intake, the challenge for the thyroid gland is to adjust the hormonal output in order to achieve the new equilibrium and maintain it until term: this corresponds to physiological adaptation of the thyroidal economy to the pregnant state. When pregnancy takes place in women who are otherwise healthy but reside in an area with a restricted iodine intake, physiological adaptation is progressively replaced by pathological alterations. Thus, pregnancy typically acts to reveal the underlying iodine restriction: the more severe the ID, the more pronounced the maternal (and also fetal) thyroidal consequences [10–15].

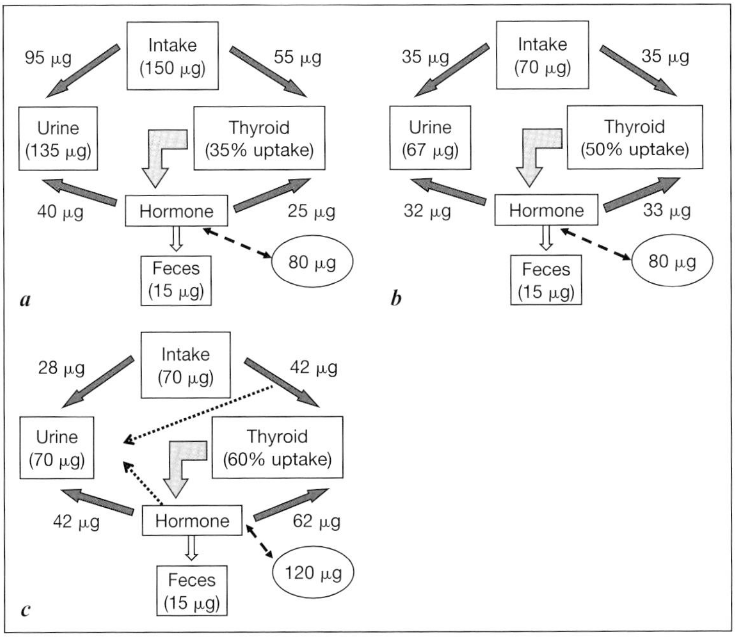

Fig. 2. Schematic representation of the kinetics of iodide in healthy nonpregnant and pregnant adults. ***a*** Nonpregnant adult with an adequate iodine intake of 150 µg/day. ***b*** Nonpregnant adult with a restricted iodine intake corresponding to 70 µg/day. ***c*** The latter condition is compared with an identically restricted level of iodine intake (70 µg/day) in a pregnant woman. Daily production of TH was set at 80 µg of iodide/day (in nonpregnant adults) and increased by 1.5-fold to 120 µg/day during pregnancy.

Metabolism of Iodine during Pregnancy

After being reduced to iodide, dietary iodine is rapidly absorbed from the gut. Iodide of dietary origin then mixes rapidly with iodide derived from the peripheral catabolism of TH and iodothyronines by deiodination, and together they constitute the extrathyroidal pool of plasma inorganic iodide. This pool exists in a dynamic equilibrium with two organs, the thyroid gland and the kidneys. A schematic representation of the kinetics of iodine in healthy nonpregnant and pregnant women is shown in figure 2. A normal adult uses some 80 µg of iodide/day to produce TH and the system is balanced to fulfil these daily needs. When the iodine intake by nonpregnant women is adequate (~150 µg/day), the kinetic balance is achieved by thyroid uptake of 35% of the available iodine. Of the 80 µg of hormonal iodide produced each day by the catabolism of TH, 15 µg of iodide is lost in the feces, leaving 65 µg to be redistributed between the thyroid gland and irreversible urinary losses. In these conditions the metabolic balance remains in equilibrium and the body is able to maintain abundant intrathyroidal iodine stores ranging from 10 to 20 mg [16]. In contrast, when the iodine intake is restricted before the onset of a pregnancy to 70 μg/day or less, the body must increase iodide trapping through the pituitary-thyroid feedback mechanism to compensate for iodine restriction and maintain the necessary absolute iodine intake. Augmentation of iodide trapping is the fundamental mechanism through which the thyroid gland adapts to changes in the daily iodine supply, and this mechanism is the key to understanding the thyroid adaptation to ID [17, 18]. In such conditions, there is a shortfall of some 10 μg of iodine a day and the thyroid gland uses stored iodine which is therefore progressively depleted to low amounts of 2–5 mg of stable iodine. Over time, if the nutritional situation remains unchanged, the metabolic balance of iodine tends to become negative.

Two fundamental changes take place during pregnancy. First, there is a significant increase in renal iodide clearance by some 30–50%. Since the renal iodide clearance already increases in the first weeks of gestation and persists thereafter, this constitutes an obligatory iodine 'leakage' which tends to lower circulating plasma inorganic iodide concentration and, in turn, induce a compensatory increase in the thyroidal clearance of iodide. Second, there is a concomitant and sustained increase in the production of TH by 50%, which corresponds to an incremental requirement from 80 to 120 μg of hormonal iodide/day. These two mechanisms underscore the increased physiological activity of the thyroid gland during the first half of pregnancy [19–23]. Calculations show that when the daily intake is restricted to only some 70 μg of iodine during pregnancy and despite an increment in thyroidal uptake to 60%, the equilibrium becomes more or less rapidly lost, with an absolute iodide entry into the gland that is insufficient to fulfil the increased requirements of TH production. In such conditions, there is a shortfall of some 20 μg of iodine/day. As figure 2 shows, in order to sustain an increased production of TH, the glandular machinery must draw from already depleted intrathyroidal iodine stores [2, 13]. An additional mechanism of maternal iodine deprivation occurs during the second half of gestation, from the passage of a part of the available iodine from the maternal circulation to the fetal-placental unit. The absolute extent of iodine that is transferred from the mother to the fetus has not yet been precisely established but, at midgestation, the fetal thyroid gland has already started to produce TH that are indispensable for the adequate development of the fetus.

In summary, during pregnancy in conditions with ID, the already lowered intrathyroidal iodine stores become even more depleted within one trimester after conception; furthermore, when the iodine deprivation prevails already during the first half of gestation, ID tends to become even more severe with the progression of gestation to its final stages. This is the rationale for the excessive stimulation of the thyroid gland that is observed during a pregnancy that takes place in conditions with ID. The consequences of this are relative hypothyroxinemia and hypothyroidism with an increased concentration in serum thyroid-stimulating hormone (TSH) and thyroglobulin (TG), and finally an increase in thyroid volume (TV) leading to goiter formation in both the mother and newborn [24–26].

Epidemiology and Management of ID during Pregnancy

Epidemiological Aspects

Countries such as the United States, Japan, and a limited number of European regions have set up national programs of dietary iodine supplementation in the population that have been in place for many years. Therefore, ID disorders are believed not to present problems. This view is however probably too optimistic. A recent survey in the United States has shown that the average iodine intake has markedly decreased during the period 1988–1994, compared with a similar survey carried out in 1971–1974. The median urinary iodine excretion is presently 150 μg/l compared with over 300 μg/l in the previous period. Even though such a level of iodine intake in the USA may at first glance be considered almost ideal and 'comfortably above the recommended minimum', the survey showed that as many as 15% of women of child-bearing age and almost 7% during pregnancy had iodine excretion levels which were in the range of moderate ID, namely below 50 μg/l [27, 28].

A second important epidemiological consideration is that the risk of iodine deprivation during pregnancy needs to be assessed locally and monitored over time, because mild to moderate ID occurs in areas that are not immediately recognized to be iodine-deficient. For instance, the southwestern region of France was not particularly known to be iodine deficient because of the relative proximity to the sea and the fish-eating habit of the population. Nevertheless, a study performed in 1997 in a cohort of pregnant women in the city of Toulouse clearly showed that urinary iodine excretion levels (UIE) were too low, with over 75% of pregnant women having iodine excretion levels below 100 μg/l [29].

A third important concept relates to the notion of unexpected geographical variations in the iodine intake within a given country, because ID in general, and mild-to-moderate ID more specifically, may frequently show significant variations from one area to another. A good example is illustrated by a Danish study. Pregnant women without iodine supplementation had a median iodine excretion level of 62 μg/g creatinine in Copenhagen compared with only 33 μg/g creatinine in another area of Denmark (Jutland). Furthermore, these striking differences were not alleviated in pregnant women from the same two areas who received iodine supplements, indicating that the supplementation was not sufficient enough, presumably because the iodine supplements were entirely taken up by the maternal (and perhaps fetal?) iodine-deprived thyroid glands [30].

In summary, the degree of ID should therefore be assessed specifically in each area concerned and the local situation correctly evaluated before embarking on medical recommendations for iodine fortification programs [31].

Managerial Aspects

In 2001, the World Health Organization has officially endorsed the recommendations made by international organizations such as the ICCIDD (International Council for Control of Iodine Deficiency Disorders) and UNICEF (United Nations Children's Fund) to eliminate ID disorders, on the basis that ID present at critical stages during pregnancy and early childhood resulted in impaired development of the brain and consequently in impaired mental function. The recommended nutrient intake (RNI) for iodine in adults and children above the age of 12 years is 150 μg/day. While a variety of methods exists for the correction of ID, the most commonly applied method is universal salt iodization (USI), that is the addition of suitable amounts of potassium iodide (or iodate) to all salt for human and livestock consumption [32]. In January 2005, a committee of international experts met in Geneva under the auspices of the World Health Organization, and the 2001 recommendations were revised [33] (table 1). The consensus reached by the panel was that the RNI for iodine during pregnancy and breastfeeding should range between 200 and 300 μg/day, with an average of 250 μg/day. During breastfeeding, the physiology of TH production returns to normal but iodine is efficiently concentrated by the mammary gland into milk. Since breast milk provides approximately 100 μg of iodine per day to the infant, the WHO recommendation is that breastfeeding mothers should continue to take 250 μg of iodine/day. An excessive iodine intake may potentially cause more disease, especially in patients with known (or underlying) autoimmune thyroid disorders or autonomous thyroid tissue [34]. Since there is no strong evidence to define clearly 'how much more iodine may become too much iodine', the present consensus is to indicate that there is no proven further benefit in providing pregnant women with more than twice the daily RNI, i.e. 500 μg of iodine/day.

To implement the RNI for iodine in the pregnant state, the natural iodine intake level in a population must be taken into account. Therefore, multiple tailored means must be used to reach the RNI for iodine. The overall consideration is that the sooner the iodine fortification is implemented, the better is the resulting adaptation of thyroid function to the pregnant state. It is also important to

Table 1. Recommendations for iodine nutrition during pregnancy

- It is recommended that women take 150 μg of iodine/day before pregnancy
- It is recommended that women increase their iodine intake to 250 μg/day during pregnancy and breastfeeding
- It is recommended to start iodine fortification as early in pregnancy as possible
- The maximum iodine intake should not exceed 500 μg/day to prevent the risk of iodine-induced thyroid disease in women with a predisposition (autoimmune thyroid disease, autonomous thyroid tissue)
- USI cannot achieve target supplementation levels because of salt restriction in pregnant women
- To achieve the RNI level for iodine supplementation, it is recommended to use:
  - Oral iodine supplements with KI (100–200 μg/day) or multivitamin tablets containing iodine (125–150 μg/day) in countries with a partial USI coverage
  - Oral iodized oil (400 mg of iodine once): this dose supplements for a 1-year period and is adequate for pregnant women who cannot afford or adhere to daily supplementation
- To monitor the adequacy of iodine supplementation at a population level, it is recommended to use measurements of urinary iodine excretion
- To monitor the adequacy of iodine supplementation at an individual level, it is recommended to use thyroid function parameters, such as serum TSH, free T_4, TG, T_3/T_4 ratio, and TV measured by ultrasonography

emphasize that USI cannot be used for this purpose in pregnancy because of the necessary salt restriction. Practically for the implementation of iodine fortification during pregnancy, several epidemiological situations must be distinguished. In countries with a long-standing and well-established USI program, pregnancies are not at risk of having ID and, therefore, no systematic dietary fortification ought to be organized in these populations. It can however be recommended individually to pregnant women to use multivitamin tablets prepared specifically for the needs of pregnancy and containing iodine supplements, since it is known that even in such apparently satisfactory iodine intake conditions, a fraction of pregnant women may still have an insufficient dietary iodine intake [27]. In countries without an efficient USI program or an established USI program where the coverage is known to be only partial, complementary approaches are required to reach the RNI for iodine. Such approaches include the use of oral iodine supplements in the form of potassium iodide (100–200 μg/day) or the inclusion of KI (125–150 μg/day) in multivitamin tablets specifically designed for pregnancy. Finally in areas with no accessible USI program and difficult socioeconomic conditions generally, it is recommended to administer orally iodized oil as early during gestation as possible: 400 mg of iodine will cover thyroidal needs for about a 1-year period [35].

Monitoring the Adequacy of Iodine Intake

The best single test to evaluate the adequacy of iodine nutrition in a population is provided by UIE. In conditions with an adequate iodine intake during pregnancy, the UIE should ideally range between 150 and 250 μg/day (or 100–200 μg/l, based on an average 1.5 liters of daily urine output) [36]. However, although UIE is highly useful for public health estimations of iodine intake levels in populations, it alone is not a valid diagnostic criterion in individuals. Therefore, to assess the situation at the level of an individual, it is recommended to use thyroid function parameters which constitute valid markers of the thyroidal consequences of ID during pregnancy (table 1).

Consequences of ID during Pregnancy

Biological Consequences

The increase in TBG during gestation causes an increase in total serum TH (T_4 and T_3). To estimate the free hormone concentration, a TH binding ratio, free T_4 index, or direct free T_4 measurement must be obtained. Because the reduction in the free fraction of T_3 is approximately equal to that of T_4, the standard approach for these determinations using T_3 as a tracer can still be used. However, it is important to recognize that as the free fraction is reduced, the resin T_3 uptake (and similar assessments of the free hormone fraction) asymptotically approaches a fixed lower limit. This is not linearly related to the increase in unoccupied TBG binding sites [37]. Thus, the decrease in the TH binding ratio usually does not match the quantitative decrease in the T_4- and T_3-free fractions estimated directly, and in some sera the free T_4 index or estimate will end up being slightly elevated relative to the true free T_4 or T_3 [38]. Direct measurements of serum free T_4 using the older 'analogue' technologies often resulted in an artifactually decreased free T_4 estimate in euthyroid pregnant subjects. Such artifacts have been attributed to the influence of the physiological serum albumin decrease that commonly occurs in pregnancy. Even nowadays and despite technical progress, these artifacts need to be taken into account in the interpretation of thyroid function tests in pregnant women [39, 40].

In healthy pregnant women with an adequate iodine supply, the increase in total serum T_4 is less marked than the increase in serum TBG. As a consequence, TBG is slightly less saturated with T_4 and the free T_4 concentration decreases physiologically during gestation by 10–15% [2]. To differentiate between the 'normal' and an abnormal decrement in free T_4 concentrations, it has recently been suggested to establish trimester-specific ranges for free T_4 measurements in the pregnant state [41]. Concerning serum TSH concentrations, these may be transiently lowered – and hence become infranormal – during

Table 2. Thyroid physiology during pregnancy

Physiological change	Thyroid-related consequences	
	in iodine sufficiency	in ID
Increased renal iodine clearance with obligatory iodine losses	Increased thyroidal clearance	Increased thyroidal clearance (more pronounced)
Decreased plasma iodine and placental transport of iodine to the fetus	Serum free T_4 decreases marginally and TSH remains unchanged	Serum free T_4 decreases significantly and TSH increases progressively until term
Increased serum TBG	Increased serum total T_4 and T_3	Increased serum total T_4 and T_3 that do not reach ideal levels
Increased plasma volume	Increased T_4 and T_3 pool size	Increased T_4 and T_3 pool size
Inner-ring deiodination of T_4 and T_3 by placenta	Accelerated rates of T_4 and T_3 degradation and production	Accelerated rates of T_4 and T_3 degradation and production

the first trimester in about one fifth of pregnant women in response to elevations of human chorionic gonadotropin. Thus, such a lowering in serum TSH should not lead automatically to a diagnosis of thyroid dysfunction. During the second and third trimester, serum TSH returns to the normal range of 0.4–2.5 mU/l [2, 5, 7, 42] (see table 2).

The main impact of ID occurring before and during pregnancy, even when it is considered only mild or moderate, is to induce maternal hypothyroxinemia. Hypothyroxinemia can be defined as any deviation of serum free T_4, at any time point during gestation, that is present for a prolonged period with free T_4 concentrations significantly below those considered ideal for a given gestational age. In these conditions, the abnormal lowering in free T_4 leads, in turn, to enhanced thyroidal stimulation via the pituitary (TSH) feedback mechanisms and ultimately to goiter formation in both the mother and fetus (see below). In clinical practice, enhanced glandular stimulation associated with ID can be evaluated using simple biochemical parameters [13]. Five biochemical thyroid parameters have been shown to provide useful markers for evaluating enhanced glandular stimulation when pregnancy takes place in association with ID. The first parameter is relative hypothyroxinemia, i.e. free T_4 concentrations that tend to cluster near (or below) the lower limit of normality. The second parameter is preferential T_3 secretion, reflected by an elevated total T_3/T_4 molar ratio.

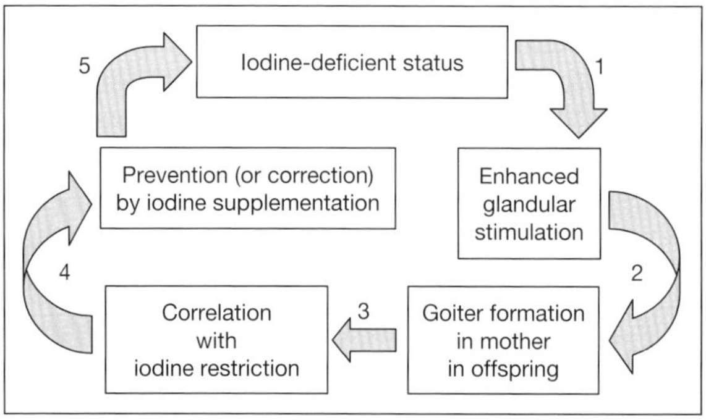

Fig. 3. Schematic representation of the formation of a vicious circle when pregnancy takes place in conditions with ID to illustrate the concept of enhanced thyroidal stimulation leading to goiter formation, and its prevention by fortification of dietary iodine intake.

The third parameter is related to changes in serum TSH. While serum TSH remains unchanged in iodine-sufficient pregnancies, TSH levels show a progressive and steady increase after the first trimester and until term in women with ID. The fourth parameter concerns the changes in serum TG. In iodine-deficient pregnancies, serum TG increases progressively to reach supranormal values, mainly during late gestation. It is important to emphasize that monitoring serum TG changes during pregnancy in conditions with ID is of particular clinical value, because the increment in TG correlates well with goiter formation, hence constituting a useful prognostic marker of gestational goitrogenesis. Finally, as already alluded to above, the fifth parameter is provided by the lowering in UIE, which broadly follow the degree of severity of ID. The successive steps in the formation of a vicious circle associated with ID, and its prevention by the early fortification of dietary iodine intake are schematically represented in figure 3.

Goiter Formation

In the early 1990s, the concept was introduced that ID was a preponderant causal factor to explain gestational goitrogenesis, affecting both the mother and the fetus (fig. 4). While goiter formation was not noticeable in pregnant women residing in iodine-sufficient areas, several European studies indicated that significant changes in TV occurred in association with pregnancy. Together, these studies have shown that pregnancy is frequently associated with goiter formation [2, 10, 24]. In regions with a sufficient iodine intake, the increments in TV remain usually minimal, in the order of 10–15% on average above the preconception TV; these minor changes are mainly consistent with vascular swelling (intumescence) of the gland during pregnancy [43, 44]. In other regions known to have a lower iodine intake, increments in TV are significantly more marked, ranging between 20 and 35% on average, with many women exhibiting a doubling in TV between the first trimester and term. For instance in areas such as Brussels (Belgium) and Toulouse (France), 10% of the women were shown to develop a gestational goiter before iodine supplementation was systematically prescribed, and the degree of glandular hyperplasia was directly correlated to the severity of ID during pregnancy (see fig. 4b) [5, 24, 29]. Thus, goiter formation during pregnancy is the hallmark of ID. Together, low intrathyroidal iodine stores that prevail already before conception, increased needs for a higher iodine availability once pregnancy begins, and finally, insufficiency of daily iodine intake – that is maintained throughout gestation – constitute the three major components of enhanced thyroidal stimulation and resulting goitrogenesis in the pregnant state [45–47].

Concerning the newborns of mothers with ID, precise ultrasonographic measurements of TV in neonates have indicated that thyroid sizes were 40% larger in the newborns from nonsupplemented mothers compared with newborns from iodine-supplemented mothers (see fig. 4e). Moreover, glandular hyperplasia was already present in 10% of these infants soon after birth, compared with none in the newborns from iodine-receiving mothers. These data show that ID is associated with goiter formation in the progeny, and emphasize the exquisite sensitivity of the fetal thyroid to the consequences of maternal iodine deprivation, indicating that the process of goiter formation starts already during the earliest stages of development of the fetal thyroid gland [48].

Long-Term Consequences of Gestational Goitrogenesis

An important question concerns the long-term evolution of a goiter formed during pregnancy, in the absence of iodine supplementation. Both prospective and retrospective studies have shown that maternal goiters due to ID do not revert entirely to baseline TV values after parturition [45, 49]. The women who develop a goiter during gestation are prone to remain goitrous thereafter, and the succession of consecutive pregnancies add to this detrimental effect (see fig. 4d). Gestational goiter formation constitutes therefore one of the environmental factors explaining the preponderance of goiter in the female population, especially in multiparous women (see fig. 4c). In a recent study from Denmark, it was also shown that parity had an influence on thyroid size in conditions with ID, especially in women who smoked (see fig. 4e) [47].

In summary, pregnancy represents a strong goitrogenic stimulus for both the mother and fetus, even in areas with only moderate ID. Several environmental factors may play a role to explain gestational goiter formation, which tend to reinforce each other: ID, successive pregnancies, and finally smoking habits.

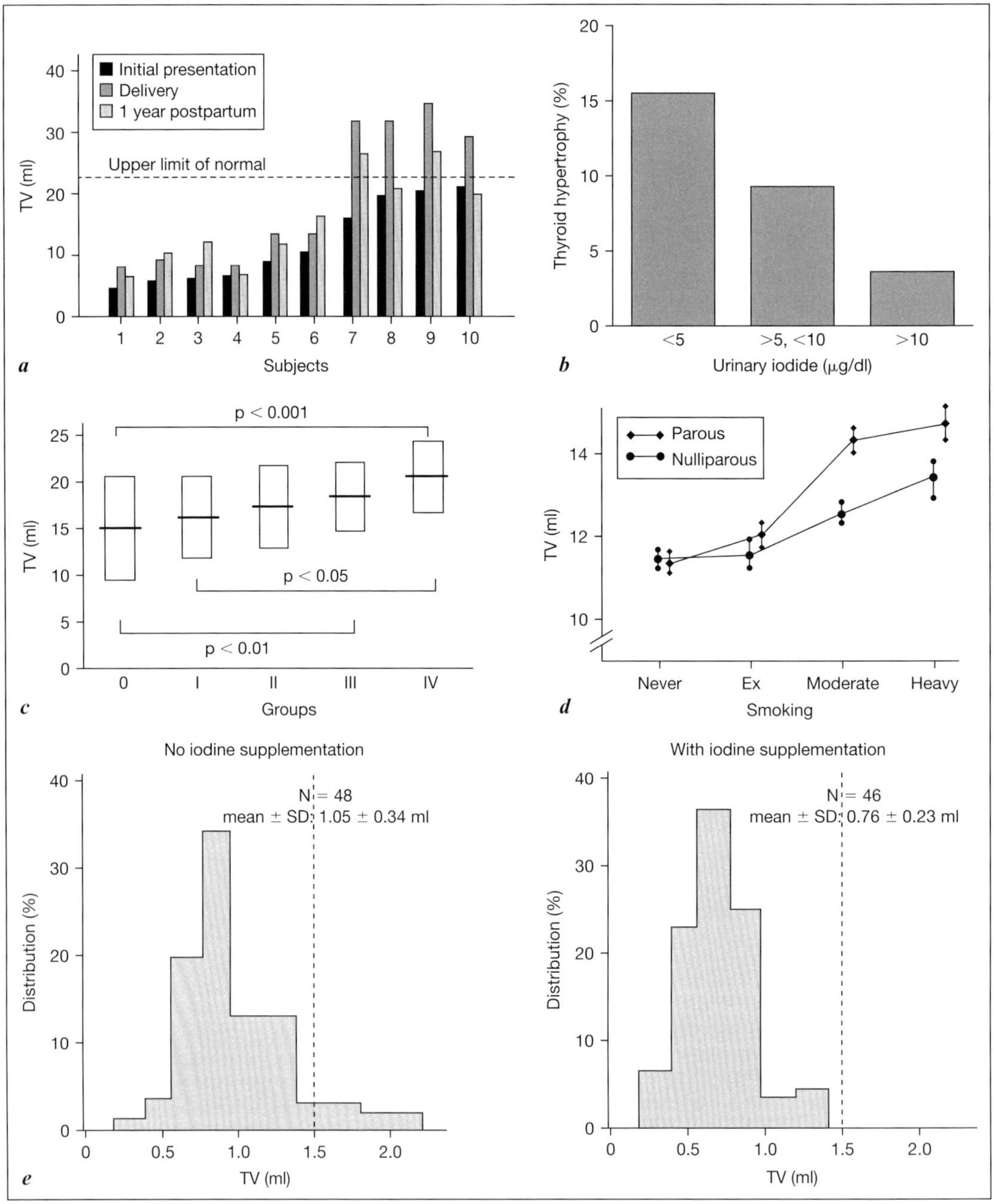
Initial presentation
Delivery
1 year postpartum
Upper limit of normal
TV (ml)
Subjects
a
Thyroid hypertrophy (%)
<5
>5, <10
>10
Urinary iodide (µg/dl)
b
p < 0.001
p < 0.05
p < 0.01
TV (ml)
Groups
0
I
II
III
IV
c
Parous
Nulliparous
TV (ml)
Never
Ex
Moderate
Heavy
Smoking
d
No iodine supplementation
N = 48
mean ± SD: 1.05 ± 0.34 ml
Distribution (%)
TV (ml)
With iodine supplementation
N = 46
mean ± SD: 0.76 ± 0.23 ml
Distribution (%)
TV (ml)
e

Prevention of Gestational Goiter Due to ID

As already discussed, women should ideally be provided with an adequate iodine intake (150 μg/day) long before they become pregnant in order to prevent gestational goitrogenesis, since it is only by reaching a long-term steady state with replenished intrathyroidal iodine stores that the triggering of the thyroid machinery can be avoided once gestation begins. To achieve such a goal, national public health authorities need to develop iodine supplementation programs of the population's diet. Correcting this public health problem has been the aim of a massive global campaign, undertaken 10–15 years ago worldwide, and that has shown remarkable progress so far [31, 50–52]. Until 1992, most European countries were moderately to severely iodine deficient. A survey carried out in twelve European countries in more recent years, using a mobile unit (the 'ThyroMobil' van) equipped with a sonographic device and the facilities for collecting urine samples, allowed for the determination of TV and urinary iodine concentrations in almost 8,000 schoolchildren aged 7–15 years. The main results indicated that the status of iodine nutrition was markedly improved in many, albeit not all, of the European countries surveyed. Silent iodine prophylaxis is not sufficient to restore an adequate iodine balance, and more stringent prophylactic measures need to be taken by public health authorities to achieve an ideal iodine nutrition status in the population [53–55].

In the mean time, the most appropriate preventive and therapeutic approach to avoid gestational goitrogenesis is to systematically increase the iodine supply as early as possible during gestation and continue the nutrition fortification after parturition, particularly in mothers who breastfeed. This can

Fig. 4. ***a*** TV in 10 selected healthy pregnant women (numbered from 1 to 10) in Brussels, measured by ultrasonography at initial presentation in the first trimester, then at delivery, and finally 12 months postpartum, and showing goiter formation during pregnancy in 4/10 women and only a partial TV normalization during the postpartum period (adapted from Glinoer et al. [45]). ***b*** Inverse correlation between the degree of thyroid hypertrophy and the severity of ID (measured as urinary iodine concentrations) in pregnant women in the southwest of France (adapted from Caron et al. [29]). ***c*** Progressive changes in TV (mean ± SD) between group 0 (representing nulliparous women) and group IV (representing women with previous pregnancies: group I had one pregnancy, group II two pregnancies, group III three pregnancies, group IV four or more pregnancies) in a retrospective study of TV in relation to parity in 208 nongoitrous healthy women with a mean age of 42 years from southern Italy (adapted from Rotondi et al. [46]). ***d*** The effect of parity and smoking on TV in women in Denmark (adapted from Knudsen et al. [47]). ***e*** Distribution frequencies of TVs in neonates born to mothers without (left) and with (right) iodine supplementation during pregnancy. Iodine supplementation allowed for a marked 38% average reduction in mean neonatal TV and for the complete prevention of thyroid hyperplasia at birth. The upper limit of normal TV in newborns is indicated by the vertical dotted lines (adapted from Glinoer et al. [48]).

be achieved by the use of multivitamin pills, containing appropriate amounts of iodine supplements. How much supplemental iodine should be given to prevent goiter formation remains a matter of local assessment, since it depends mainly on the extent of the preexisting iodine deprivation [56–59]. The ultimate goal which is to restore and maintain a balanced iodine status can be reached in most instances with 100–200 μg iodine given daily as a supplement during pregnancy. It should be remembered, however, that with long-standing iodine restriction in the diet before the onset of pregnancy, a lag period (of about one trimester) is inevitable before the benefits of the iodine supplementation improving thyroid function are observed.

Pregnancy in Regions with Severe ID

In many regions in the world, ID is not only overtly present, but it is often severe. Large areas remain in Central Africa and Asia, for instance, that still have iodine intakes below 25 μg/day, characteristic of severe ID and endemic goiter [51, 52]. In such regions, the thyroid status of pregnant women and their offspring is frequently impaired. In addition, other factors, such as selenium deficiency and thiocyanate excess (resulting from the staple use of foodstuffs such as cassava) combine with severe ID, tending to complicate the thyroidal situation even further [60]. In terms of thyroid function, the adult populations usually exhibit a mixed pattern encompassing subjects with a normal thyroid function and others who present various degrees of hypothyroidism. In women of child-bearing age, severe ID and hypothyroidism play a role in reducing fertility and increase the rate of spontaneous abortions.

When these women become pregnant, thyroid function tends to deteriorate even further, as gestation progresses. Thus, the thyroidal stress associated with pregnancy in conditions with mild to moderate ID cannot be compared, at least in quantitative terms, with the thyroidal repercussions observed in countries with long-standing and severe ID. Because of the obvious difficulties inherent in careful field studies in most areas with severe ID, there have been only few studies of thyroid function and no systematic study to assess the changes in goiter size in pregnant women [61–63]. Until a few years ago, it was not feasible to obtain echographic measurements of the thyroid gland on a large and representative scale; it was even more difficult to investigate prospectively goitrogenic changes during pregnancy. Presently, this situation is rapidly evolving, because of the possibility to adapt the ThyroMobil technology to large field studies even in remote areas [64, 65].

In women of child-bearing age and during pregnancy, iodine supplements have been administered in the form of iodized salt, potassium iodide drops, and also in the form of iodized oil (given intramuscularly or orally) as an emergency prophylactic and therapeutic approach in areas with severe ID complicated by endemic cretinism. Several such programs have conclusively demonstrated their remarkable efficiency to treat endemic goiter, as well as to eradicate endemic cretinism [31, 60, 66]. Also, the results of these studies have proved that the pregnancy of women who reside in severely iodine-deficient regions can be managed adequately with iodine supplementation. Except for emergency situations, there is presumably no need to use supraphysiological amounts of iodine to improve significantly or even normalize thyroid function parameters. Although it has not been possible so far in the setting of difficult field studies to evaluate quantitatively the reduction in goiter size or goiter prevalence associated with the improvement of thyroid function, it can be assumed that goiter reduction was indeed a 'side' benefit of the improvement in the iodine status.

Fetal and Neonatal Consequences of Maternal ID and Thyroid Underfunction

The adequate functioning of both the maternal and fetal thyroid glands plays an important part to ensure that the fetal neuropsychointellectual development progresses normally. Globally, three sets of clinical disorders, schematically illustrated in figure 5, ought to be considered. In infants with a defect of thyroid ontogenesis, leading to congenital hypothyroidism, the participation of maternal TH to the fetal circulating thyroxine environment remains theoretically normal. Therefore, the risk of brain damage results exclusively from the insufficient production of TH by the fetus, and will hence depend on the severity of the defect involved (for instance, total agenesis versus ectopia). In contrast, when only the maternal thyroid gland is functionally deficient (for instance, in hypothyroidism due to chronic autoimmune thyroiditis), both the severity and the temporal occurrence of maternal thyroid underfunction will drive the resulting consequences for an impaired fetal neuronal development. Clinical situations of this type may take place already at early gestational stages (women with untreated or undertreated hypothyroidism), or appear during late gestational stages (women with autoimmunity features who are euthyroid during the first half of gestation). Finally, in conditions of ID, both maternal and fetal thyroid functions are affected. Therefore, it is primarily the precocity and severity of pregnancy-related ID that will drive the potential repercussions for fetal neurological development.

Iodine is required for the synthesis of TH, and TH are crucial for brain development both during fetal and early postnatal life [67–70]. When severe enough, ID induces maternal and fetal hypothyroxinemia from early gestation

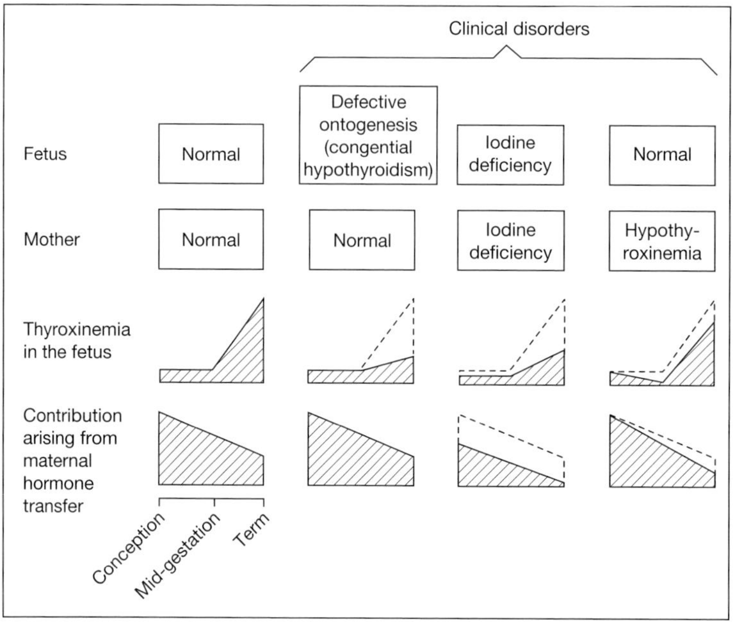

Fig. 5. Schematic representation of the three sets of clinical conditions that may affect thyroid function in the mother alone, the fetus alone, or the fetomaternal unit, showing the relative contributions of an impaired maternal and/or fetal thyroid function that may eventually lead to alterations in fetal thyroxinemia (from Glinoer and Delange [25]).

onwards [25]. Thus, any impairment in hormone availability during critical periods of brain development may induce irreversible brain damage, with mental retardation and neurological abnormalities as the final consequences (i.e. endemic cretinism) that ultimately depend upon the timing and severity of the brain's insult [71–73]. The characteristic neurological picture of endemic cretinism is presumably directly due to insults to the developing brain, occurring already during the first trimester (i.e. deafness) and mostly during the second trimester, while cerebellar abnormalities may result from a postnatal insult [74–76]. This interpretation is supported by the observation that the full neurological picture of endemic cretinism can only be prevented when ID is corrected before the second trimester of pregnancy, and optimally prior to conception [71]. The particular pattern of myxedematous cretinism that is commonly found in Africa might be explained by the fact that in this area severe ID is complicated by selenium deficiency. Selenium deficiency results in the accumulation of peroxide in the hyperstimulated thyroid glands, and excess peroxide induces thyroid cell destruction, leading to parenchymal fibrosis and hypothyroidism [77]. Endemic

Table 3. Neuropsychointellectual deficits in schoolchildren born to mothers residing in areas with mild-moderate ID

Region	Tests	Findings	Reference
Spain	Locally adapted tests: Bayley McCarthy Cattell	Lower psychomotor and mental development	Bleichrodt et al. [79]
Italy Sicily	Bender-Gestalt	Low perceptual integrative motor ability and neuromuscular and neurosensorial abnormalities	Vermiglio et al. [80]
Italy Tuscany	Wechsler Raven	Low verbal IQ, perception, motor and attentive functions	Fenzi et al. [81]
Italy Tuscany	WISC Reaction time	Lower velocity of motor response to visual stimuli	Vitti et al. [82] Aghini-Lombardi et al. [83]
Italy Sicily	DSM-IV-TR, validated by subscales for the Italian population	Attention deficit and hyperactivity disorder	Vermiglio et al. [84]

Modified from Glinoer and Delange [25].

cretinism, therefore, constitutes the extreme expression of a spectrum of abnormalities in the physical and intellectual development in children, as well as diminished functional capacity of the thyroid gland, observed in inhabitants of areas with severe ID and endemic goiter. In a meta-analysis of eighteen studies conducted in areas with severe ID, it was shown that ID was responsible for an IQ loss of 13.5 points [78].

Neurointellectual deficits associated with ID, however, are not limited to remote areas, known to be severely affected by ID. In recent years, the impact of mild or moderate ID on the fetus has also been recognized. For instance in studies conducted in areas with only a moderate or even mild ID (such as in the southern part of Europe), developmental abnormalities were shown to occur in clinically euthyroid school-age children [79–84] (table 3). Thus, ID is one of the most prevalent causes of mental retardation and reduced learning abilities in children that could easily be prevented and eliminated by adequate iodine supplementation.

Table 4. Take home messages

- The challenge for the thyroid gland is to adjust the hormonal output to achieve a new steady state corresponding to physiological adaptation to the pregnant state. When pregnancy takes place in women with a restricted iodine intake, physiological adaptation is progressively replaced by pathological alterations
- The RNI for iodine is 250 μg/day in pregnant and lactating women
- The most important consequence of ID during pregnancy is maternal underfunction (i.e. hypothyroxinemia), leading in turn to excessive thyroidal stimulation with a rise in serum TSH and increased serum TG
- Gestational goiter formation is the clinical hallmark of ID. Together, low intrathyroidal iodine stores before conception, increased needs for higher iodine availability once pregnancy begins, and the insufficiency of daily iodine intake constitute the major components of enhanced thyroidal stimulation leading to goitrogenesis affecting both the mother and fetus
- A goiter formed during gestation regresses only partially after parturition. Pregnancy is therefore an environmental factor that may help explain the higher prevalence of goiter and thyroid disorders in women. Goiter formation takes place in the fetus, emphasizing the exquisite sensitivity of the fetal thyroid to consequences of maternal iodine deprivation, and indicating that the process of goiter formation begins during the earliest stages of fetal gland development
- Iodine is required for the synthesis of TH, and TH are crucial for fetal brain development. When severe enough, ID may induce fetal hypothyroxinemia from early gestation onwards. Any impairment in hormone availability during critical periods of development may induce brain damage, with irreversible neurological abnormalities
- In recent years, the impact of mild, moderate, or only borderline ID on the fetus has been recognized with developmental abnormalities shown to occur and impaired school achievements that may be present in apparently normal school-aged children

Conclusions and Perspectives

The main changes in thyroid function associated with the pregnant state are due to increased hormone requirements, which begin in the first trimester of gestation. Increased hormone requirements can only be met by proportionally increased hormone production, directly depending upon the availability of iodine in the diet. When dietary iodine is deficient, adequate physiological adaptation is difficult to achieve and is progressively replaced by pathological alterations occurring in parallel with the degree of long-term iodine deprivation, leading to enhanced glandular stimulation. Therefore, pregnancy typically acts to reveal the underlying iodine restriction and gestation results in a deficient iodine status, even in conditions with only a marginally restricted intake, such as is observed in many European regions. ID during pregnancy has important repercussions for both mother and fetus, namely thyroid underfunction and goitrogenesis. Furthermore, ID may be associated with alterations of the neuropsychointellectual outcome in the progeny, and the risk for an abnormal development of the progeny is further enhanced because both the mother and offspring are exposed to the deficiency, not only during the entirety of gestation but also the postnatal period. Iodine prophylaxis should be introduced systematically to women during pregnancy and the lactation period. Concerning areas with a severe deficiency, the correction of the lack of iodine has proved highly beneficial to prevent mental deficiency disorders: the many actions undertaken to eradicate ID have prevented the occurrence of mental retardation in millions of young infants throughout the world. In most public health programs dealing with the correction of the ID disorder, iodized salt has been used as the preferred method of conveying iodine supplements to the household. Iodized salt, however, is not the ideal vector in the specific instance of pregnancy and breast-feeding, because of the necessity to restrict salt intake. Finally, it is with some concern that the results of the recent nutritional survey in the United States have disclosed that ID, thought to have been eradicated many years ago, may actually show a risk of a resurgence, particularly in young women. This issue will need to be considered seriously by the medical community and public health authorities, since similar situations may occur in other countries as well (see the main 'take home messages' in table 4).

Acknowledgments

The author acknowledgs the financial support of the Belgian Ministry for Education and Scientific Research, within the framework of the 'Actions de Recherche Concertée 2004–2009' (ARC/Convention No. 04/09-314), which was of great help in preparing the present manuscript.

References

1 Andersson M, Takkouche B, Egli I, Allen HE, de Benoist B: Current global iodine status and progress over the last decade towards the elimination of iodine deficiency. Bull WHO 2005;83: 518–525.

2 Glinoer D: The regulation of thyroid function in pregnancy: pathways of endocrine adaptation from physiology to pathology. Endocr Rev 1997;18:404–433.

3 Bartalena L: Recent achievements in studies on thyroid hormone-binding proteins. Endocr Rev 1990;11:47–64.

4 Dowling JT, Appleton WG, Nicoloff JT: Thyroxine turnover during human pregnancy. J Clin Endocrinol Metab 1964;27:1749–1750.

5 Glinoer D, De Nayer P, Bourdoux P, Lemone M, Robyn C, Van Steirteghem A, Kinthaert J, Lejeune B: Regulation of maternal thyroid function during pregnancy. J Clin Endocrinol Metab 1990;71:276–287.

6 Guillaume J, Schussler GC, Goldman J: Components of the total serum thyroid hormone concentrations during pregnancy: high free thyroxine and blunted thyrotropin (TSH) response to TSH-releasing hormone in the first trimester. J Clin Endocrinol Metab 1985;60:678–684.
7 Glinoer D, De Nayer P, Robyn C, Lejeune B, Kinthaert J, Meuris S: Serum levels of intact human chorionic gonadotropin (hCG) and its free alpha and beta subunits, in relation to maternal thyroid stimulation during normal pregnancy. J Endocrinol Invest 1993;16:881–888.
8 Bianco AC, Salvatore D, Gereben B, Berry MJ, Larsen PR: Biochemistry, cellular and molecular biology, and physiological roles of iodothyronine selenodeiodinases. Endocr Rev 2002;23:38–89.
9 Huang SA, Dorfman DM, Genest DR, Salvatore D, Larsen PR: Type 3 iodothyronine deiodinase is highly expressed in the human uteroplacental unit and in fetal epithelium. J Clin Endocrinol Metab 2003;88:1384–1388.
10 Glinoer D: Maternal and fetal impact of chronic iodine deficiency. Clin Obstet Gynecol 1997;40: 102–116.
11 Glinoer D: What happens to the normal thyroid during pregnancy? Thyroid 1999;9:631–635.
12 Glinoer D: Pregnancy and iodine. Thyroid 2001;11:471–481.
13 Glinoer D: The regulation of thyroid function during normal pregnancy – importance of the iodine nutrition status. Best Pract Res Clin Endocrinol Metab 2004;18:133–152.
14 Glinoer D: Iodine nutrition requirements during pregnancy (editorial). Thyroid 2006;16:947–948.
15 Caron P, Glinoer D, Lecomte P, Orgiazzi J, Wémeau J-L: Apport iodé en France: prévention de la carence iodée au cours de la grossesse et l'allaitement. Ann Endocrinol 2006;67:281–286.
16 Delange F, Bourdoux P, Chanoine JP, Ermans AM: Physiopathology of iodine nutrition during pregnancy, lactation, and early postnatal life; in Berger H (ed): Vitamins and Minerals in Pregnancy. New York, Raven Press, 1988, pp 205–214.
17 Ermans AM: Endemic goiter; in Ingbar SH, Braverman LE (eds): The Thyroid – a Fundamental and Clinical Text, ed 5. Philadelphia, Lippincott, 1986, pp 705–721.
18 Robbins J: Iodine deficiency, iodine excess and the use of iodine for protection against radioactive iodine. Thyroid Today 1994;3:1–5.
19 Pochin EE: The iodine uptake of the human thyroid throughout the menstrual cycle and in pregnancy. Clin Sci 1952;11:441–445.
20 Halnan KE: The radioiodine uptake of the human thyroid in pregnancy. Clin Sci 1958;17: 281–290.
21 Aboul-Khair SA, Crooks J, Turnbull AC, Hytten FE: The physiological changes in thyroid function during pregnancy. Clin Sci 1964;27:195–207.
22 Silva JE: Effects of iodine and iodine-containing compounds on thyroid function. Med Clin North Am 1985;69:881–898.
23 Liberman CS, Pino SC, Fang SL, Braverman LE, Emerson CH: Circulating iodide concentrations during and after pregnancy. J Clin Endocrinol Metab 1998;83:3545–3549.
24 Glinoer D, Lemone M: Goiter and pregnancy – a new insight into an old problem. Thyroid 1992;2: 65–72.
25 Glinoer D, Delange F: The potential repercussions of maternal, fetal, and neonatal hypothyroxinemia on the progeny. Thyroid 2000;10:871–887.
26 Glinoer D: The importance of iodine nutrition during pregnancy. Public Health Nutrition April–May 2007, in press.
27 Hollowell JG, Staehling NW, Hannon WH, Flanders DW, Gunter EW, Maberly GF, Braverman LE, Pino S, Miller DT, Garbe PL, DeLozier DM, Jacjson RJ: Iodine nutrition in the United States. Trends and public health implications: iodine excretion data from national health and nutrition examination surveys I and III (1971–1974 and 1988–1994). J Clin Endocrinol Metab 1998;83: 3401–3408.
28 The Public Health Committee of the American Thyroid Association: Iodine supplementation for pregnancy and lactation – United States and Canada: Recommendations of the American Thyroid Association. Thyroid 2006;16:949–951.
29 Caron P, Hoff M, Bazzi S, Dufor A, Faure G, Ghandour I, Lauzu P, Lucas Y, Maraval D, Mignot F, Ressigeac P, Vertongen F, Grangé V: Urinary iodine excretion during normal pregnancy in healthy women living in the southwest of France: correlation with maternal thyroid parameters. Thyroid 1997;7:749–754.

30 Nohr SB, Laurberg P, Borlum K-G, Pedersen KM, Johannesen PL, Damm P, Fuglsang E, Johansen A: Iodine deficiency in pregnancy in Denmark – regional variations and frequency of individual iodine supplementation. Acta Obstet Gynecol Scand 1993;72:350–353.
31 Dunn JT, Delange F: Damaged reproduction: the most important consequence of iodine deficiency. J Clin Endocrinol Metab 2001;86:2360–2363.
32 World Health Organization: Assessment of Iodine Deficiency Disorders and Monitoring Their Elimination: a Guide for Programme Managers, ed 2. Geneva, WHO Press, 2001.
33 World Health Organization. Technical consultation of experts in Geneva (January 2005): The prevention and control of iodine deficiency in pregnant and lactating women and in children under two years of age – Recommendations of a WHO Technical Consultation. Public Health Nutrition April–May 2007, in press.
34 Roti E, Vagenakis AG: Effect of excess iodide – clinical aspects; in Braverman LE, Utiger RD (eds): The Thyroid – a Fundamental and Clinical Text, ed 9. Philadelphia, Lippincott Williams & Wilkins, 2005, pp 288–305.
35 Chaouki ML, Benmiloud M: Prevention of iodine deficiency disorders by oral administration of lipiodol during pregnancy. Eur J Endocrinol 1994;130:547–551.
36 Delange F: Optimal iodine nutrition during pregnancy, lactation and the neonatal period. Int J Endocrinol Metab 2004;2:1–12.
37 Wang R, Nelson JC, Weiss RM, Wilcox RB: Accuracy of free thyroxine measurements across natural ranges of thyroxine binding to serum proteins. Thyroid 2000;10:31–39.
38 Osathanondh R, Tulchinsky D, Chopra IJ: Total and free thyroxine and triiodothyronine in normal and complicated pregnancy. J Clin Endocrinol Metab 1976;42:98–103.
39 Roti E, Gardini E, Minelli R, Bianconi L, Flisi M: Thyroid function evaluation by different commercially available free thyroid hormone measurement kits in term pregnant women and their newborns. J Endocrinol Invest 1991;14:1–9.
40 Sapin R, D'Herbomez M, Schlienger JL: Free thyroxine measured with equilibrium dialysis and nine immunoassays decreases in late pregnancy. Clin Lab 2004;50:581–584.
41 Demers LM, Spencer CA: Laboratory medicine practice guidelines: laboratory support for the diagnosis and monitoring of thyroid disease. Clin Endocrinol (Oxf) 2003;58:138–140.
42 Haddow JE, Knight GJ, Palomaki GE, McClain MR, Pulkkinen AJ: The reference range and within-person variability of thyroid stimulating hormone during the first and second trimesters of pregnancy. J Med Screen 2004;11:170–174.
43 Levy RP, Newman DM, Rejali LS, Barford DA: The myth of goiter in pregnancy. Am J Obstet Gynecol 1980;137:701–703.
44 Nelson M, Wickus GG, Caplan RH, Beguin EA: Thyroid gland size in pregnancy: an ultrasound and clinical study. J Reprod Med 1987;32:888–890.
45 Glinoer D, Lemone M, Bourdoux P, De Nayer P, Delange F, Kinthaert J, Lejeune B: Partial reversibility during late postpartum of thyroid abnormalities associated with pregnancy. J Clin Endocrinol Metab 1992;74:453–457.
46 Rotondi M, Amato G, Biondi B, Mazziotti G, Del Buono A, Nicchio MR, Balzano S, Bellastella A, Glinoer D, Carella C: Parity as a thyroid size-determining factor in areas with moderate iodine deficiency. J Clin Endocrinol Metab 2000;85:4534–4537.
47 Knudsen N, Bülow I, Laurberg P, Ovesen L, Perrild H, Jorgensen T: Parity is associated with increased thyroid volume solely among smokers in an area with moderate to mild iodine deficiency. Eur J Endocrinol 2002;146:39–43.
48 Glinoer D, Delange F, Laboureur I, De Nayer P, Lejeune B, Kinthaert J, Bourdoux P: Maternal and neonatal thyroid function at birth in an area with marginally low iodine intake. J Clin Endocrinol Metab 1992;75:800–805.
49 Glinoer D, De Nayer P, Delange F, Lemone M, Toppet V, Spehl M, Grün JP, Kinthaert J, Lejeune B: A randomized trial for the treatment of mild iodine deficiency during pregnancy: maternal and neonatal effects. J Clin Endocrinol Metab 1995;80:258–269.
50 Delange F, Lecomte P: Iodine supplementation – benefits outweigh risks. Drug Safety 2000;2: 89–96.
51 Dunn JT: Correcting iodine deficiency is more than just spreading around a lot of iodine. Thyroid 2001;11:363–364.

52 Hetzel BS: Eliminating iodine deficiency disorders – the role of the International Council in the global partnership. Bull WHO 2002;80:410–417.
53 Delange F, Benker G, Caron P, Eber O, Ott P, Peter F, Podoba J, Simescu M, Szybinsky Z, Vertongen F, Vitti P, Wiersinga W, Zamrazil V: Thyroid volume and urinary iodine in European schoolchildren: standardization of values for assessment of iodine deficiency. Eur J Endocrinol 1997;136:180–187.
54 Delange F, Van Onderbergen A, Shabana W, Vandemeulebroecke E, Vertongen F, Gnat D, Dramaix M: Silent iodine prophylaxis in western Europe only partly corrects iodine deficiency: the case of Belgium. Eur J Endocrinol 2000;143:189–196.
55 Ciardelli R, Haumont D, Gnat D, Vertongen F, Delange F: The nutritional iodine supply of Belgian neonates is still insufficient. Eur J Pediatr 2002;161:519–523.
56 Romano R, Jannini EA, Pepe M, Grimaldi A, Olivieri M, Spennati P, Cappa F, D'Armiento M: The effects of iodoprophylaxis on thyroid size during pregnancy. Am J Obstet Gynecol 1991;164: 482–485.
57 Kung AWC, Lao TT, Chau MT, Tam SC, Low LC: Goitrogenesis during pregnancy and neonatal hypothyroxinemia in a borderline iodine sufficient area. Clin Endocrinol 2000;53:725–731.
58 Antonangeli L, Maccherini D, Cavaliere R, Di Giulio C, Reinhardt B, Pinchera A, Aghini-Lombardi F: Comparison of two different doses of iodide in the prevention of gestational goiter in marginal iodine deficiency: a longitudinal study. Eur J Endocrinol 2002;147:29–34.
59 Glinoer D: Feto-maternal repercussions of iodine deficiency during pregnancy. Ann Endocrinol (Paris) 2003;64:37–44.
60 Delange F: The disorders induced by iodine deficiency. Thyroid 1994;4:107–128.
61 Thilly CH, Delange F, Lagasse R, Bourdoux P, Ramioul L, Berquist H, Ermans AM: Fetal hypothyroidism and maternal thyroid status in severe endemic goiter. J Clin Endocrinol Metab 1978;47:354–360.
62 Silva JE, Silva S: Interrelationships among thyroxine, triiodothyronine, reverse triiodothyronine, and thyroid-stimulating hormone in iodine-deficient pregnant women and their offspring: effects of iodine supplementation. J Clin Endocrinol Metab 1981;52:671–677.
63 Boyages SC, Halpern JP, Maberly GF, Eastman CJ, Morris J, Collins J, Jupp JJ, Jin CE, Wang ZH, You CY: A comparative study of neurological and myxedemetous endemic cretinism in Western China. J Clin Endocrinol Metab 1988;67:1262–1271.
64 Djokomoeljanto R, Setyawan H, Dramaix M, Hadisaputro S: The thyromobil model for standardized evaluation of iodine deficiency disorder control in Indonesia. Thyroid 2001;11:365–372.
65 Rossi AC, Tomimori E, Camargo R, Medeiros-Neto G: Searching for iodine deficiency in schoolchildren from Brazil: the Thyromobil project. Thyroid 2001;11:661–663.
66 Dunn JT: The use of iodized oil and other alternatives for the elimination of iodine deficiency disorders; in Hetzel BS, Pandav CS (eds): The Conquest of Iodine Deficiency Disorders. New Delhi, Oxford University Press Publications, 1994, pp 108–117.
67 Contempré B, Jauniaux E, Calvo R, Jurkovic D, Campbell S, Morreale de Escobar G: Detection of thyroid hormones in human embryonic cavities during the first trimester of pregnancy. J Clin Endocrinol Metab 1993;77:1719–1722.
68 Chan S, Kilby MD: Thyroid hormone and central nervous system development. J Endocrinol 2000;165:1–8.
69 Lavado-Autric R, Auso E, Garcia-Velasco JV, Arufe MC, Escobar del Rey F, Berbel P, Morreale de Escobar G: Early maternal hypothyroxinemia alters histogenensis and cerebral cortex cytoarchitecture of the progeny. J Clin Invest 2003;111:1073–1082.
70 Zoeller RT: Transplacental thyroxine and brain development. J Clin Invest 2003;111:954–957.
71 Pharoah PO, Butterfield IH, Hetzel BS: Neurological damage to the fetus resulting from severe iodine deficiency during pregnancy. Lancet 1971;i:308–310.
72 Morreale de Escobar G, Obregon MJ, Escobar del Rey F: Is neuropsychological development related to maternal hypothyroidism or to maternal hypothyroxinemia? J Clin Endocrinol Metab 2000;85:3975–3987.
73 Calvo RM, Jauniaux E, Gulbis B, Asuncion M, Gervy C, Contempre B, Morreale de Escobar G: Fetal tissues are exposed to biologically relevant free thyroxine concentrations during early phases of development. J Clin Endocrinol Metab 2002;87:1768–1777.

74 Delange F: Endemic cretinism; in Braverman LE, Utiger RD (eds): The Thyroid – a Fundamental and Clinical Text, ed 8. Philadelphia, Lippincott Williams & Wilkins, 2000, pp 743–754.
75 Halpern JP, Boyages SC, Maeberly GF, Collins JK, Eastman CJ, Morris J: The neurology of endemic cretinism. Brain 1991;114:825–841.
76 DeLong GR, Stanbury JB, Fierro-Benitez R: Neurological signs in congenital iodine deficiency disorder (endemic cretinism). Dev Med Child Neurol 1985;27:317–324.
77 Vanderpas JB, Contempré B, Duale NL, Goossens B, Bebe N, Thorpe R, Ntambue K, Dumont J, Thilly CH, Diplock AT: Iodine and selenium deficiency associated with cretinism in Northern Zaire. Am J Clin Nutr 1990;52:1087–1093.
78 Bleichrodt N, Born MP: A meta-analysis of research on iodine and its relationship to cognitive development; in Stanbury JB (ed): The Damaged Brain of Iodine Deficiency. New York, Cognizant Communication Corporation, 1994.
79 Bleichrodt N, Escobar del Rey F, Morreale de Escobar G, Garcia I: Iodine deficiency – implications for mental and psychomotor development in children; in DeLong GR, Robbins G, Condliffe PG (eds): Iodine and the Brain. New York, Plenum Press, 1989, pp 269–287.
80 Vermiglio F, Sidoti M, Finocchiaro MD, Battiato S, Lo Presti VP, Benvenga S, Trimarchi F: Defective neuromotor and cognitive ability in iodine-deficient schoolchildren of an endemic goiter region in Sicily. J Clin Endocrinol Metab 1990;70:379–384.
81 Fenzi GF, Giusti LF, Aghini-Lombardi F, Bartalena L, Marcocci C, Santini F, Bargagna S, Brizzolara D, Ferretti G, Falciglia G, Montelone L, Marcheschi M, Pinchera A: Neuropsychological assessment in schoolchildren from an area of moderate iodine deficiency. J Endocrinol Invest 1990;13:427–431.
82 Vitti P, Aghini-Lombardi F, Antonangeli L, Rago T, Chiovato L, Pinchera A, Marcheschi M, Bargagna S, Bertuccelli B, Ferretti G, Sbrana B: Mild iodine deficiency in fetal/neonatal life and neuropsychological performances. Acta Med Austr 1992;19:57–59.
83 Aghini-Lombardi F, Pinchera A, Antonangeli L, Rago T, Chiovato L, Bargagna S, Bertucelli B, Ferretti G, Sbrana B, Marcheschi M, Vitti P: Mild iodine deficiency during fetal/neonatal life and neuropsychological impairment in Tuscany. J Endocrinol Invest 1995;18:57–62.
84 Vermiglio F, Lo Presti VP, Moleti M, Sidoti G, Tortorella G, Scaffidi G, Castagna MG, Mattina F, Violi MA, Crisa A, Artemisia A, Trimarchi F: Attention deficit and hyperactivity disorders in the offspring of mothers exposed to mild-moderate iodine deficiency: a possible novel iodine deficiency disorder in developed countries. J Clin Endocrinol Metab 2004;89:6054–6060.

Prof. Daniel Glinoer
Division of Endocrinology, Department of Internal Medicine
Thyroid Investigation Clinic, University Hospital Saint Pierre
322, Rue Haute, BE–1000 Brussels (Belgium)
Tel. +32 2 535 4516, Fax +32 2 535 3137
E-Mail dglinoer@ulb.ac.be or daniel_glinoer@stpierre-bru.be

Van Vliet G, Polak M (eds): Thyroid Gland Development and Function.
Endocr Dev. Basel, Karger, 2007, vol 10, pp 86–98

Ontogenesis of Thyroid Function and Interactions with Maternal Function

M.J. Obregon, R.M. Calvo, F. Escobar del Rey, G. Morreale de Escobar

Instituto de Investigaciones Biomedicas, Centro mixto 'Alberto Sols' (CSIC-UAM), Madrid, Spain

Abstract

Fetal and neonatal development of thyroid function involves the embryogenesis, differentiation and maturation of the thyroid gland, of the hypothalamic-pituitary-thyroid axis and of the systems controlling thyroid hormone metabolism. We focus here on aspects related to neurodevelopment. Throughout gestation, thyroxine (T_4) transferred from the mother, present in embryonic fluids by 4 weeks, protects the fetal brain. Free T_4 (FT_4) in fetal fluids increases rapidly, approaching adult levels by midgestation, in concentrations that are determined by the maternal serum T_4. T_3 remains very low throughout pregnancy. In the cerebral cortex T_3, generated from T_4, reaches adult values by midgestation and is partly bound to specific nuclear receptor isoforms. The iodothyronine deiodinases are important for the spatial and temporal presence of T_3 in different fetal brain areas. After onset of fetal thyroid secretion at midgestation, maternal transfer of T_4 continues to contribute importantly to fetal serum T_4, protecting neurodevelopment until birth. In rats, even a transient period of maternal hypothyroxinemia disrupts neurodevelopment irreversibly, supporting epidemiological evidence for its negative role in human neurodevelopment. The prompt treatment of maternal hypothyroidism or hypothyroxinemia should mitigate negative effects on neurodevelopment. Neurodevelopmental deficits of preterm infants might also result from an untimely interruption of the maternal transfer of T_4 [Morreale de Escobar et al: J Clin Endocrinol Metab 2000;85:3975–3987; Best Pract Res Clin Endocrinol Metab 2004;18:225–248; Eur J Endocrinol 2004;151(suppl 3):U25–U37].

Ontogenesis of Thyroid Function

The thyroid gland develops mostly during fetal and early postnatal life. The thyrocytes forming the functional unit of the gland – the thyroid follicle – derive from the embryonic endoderm in the floor of the primitive pharynx, forming a bud already visible at embryonic day 16 (E16) in humans. After proliferation of the cells and migration at E24–E32 in humans, the thyroid reaches its final position at E40–E50 [1]. An error during morphogenesis results in thyroid abnormalities, such as agenesis or ectopies, as shown in null mice with targeted deletions of thyroid transcription factors, fully described in the chapter by De Felice and Di Lauro [pp. 1–14].

The histological differentiation of the thyrocytes and formation of the follicle are accompanied by the progressive appearance of specific proteins: thyroglobulin, thyroid peroxidase, sodium/iodine symporter (NIS) and TSH receptor, all necessary for the synthesis and secretion of T_4 and T_3. The main physiological regulator of thyroid gland activity is TSH, but TSH-independent autoregulatory mechanisms also play an important role in the postnatal adaptation to fluctuations in iodine availability. Thyroglobulin has been detected in human thyroid as early as the 5th week of gestation, before the gland reaches its final position. Under the artificial conditions of organ culture, iodine uptake starts at 12–13 weeks after conception, coinciding with the closure of follicles. In vivo, however, significant uptake of iodine, a prerequisite for the synthesis and secretion of fetal thyroid hormones, is minimal until midgestation (18–20 weeks) coinciding with full development of the pituitary-portal vascular system.

For obvious reasons, studies in human fetuses are scarce and much of our knowledge of the role of thyroid hormones during fetal life comes from experimental work performed in rats [table 3 in 2; 3–5], where thyroid hormone secretion starts at E17.5–E18. Fetal serum T_4 concentrations increase 10-fold between E18 and the end of gestation (E22), a developmental period comparable to that of a human fetus at the beginning of the third trimester. T_3, however, is very low throughout gestation.

In most fetal tissues, T_4 increases in parallel to the increases in fetal plasma T_4 while T_3 concentrations are highly variable in different tissues: in liver, as in plasma, T_3 is low throughout gestation, but in other tissues, such as brain or brown adipose tissue (BAT), T_3 almost reaches adult levels. This suggests that T_3 is required for the development and maturation of these tissues, as confirmed for certain brain areas or in BAT, which is being prepared for the transition, at birth, to lower environmental temperatures. The higher T_3 observed in some tissues is due to ontogenic increases in 5′-deiodinase activities: D2 in fetal brain and BAT and D1 in lung, aimed at the production of T_3 when and where T_3 is required [3, 5]. The presence of 5-deiodinase (D3) activity is high during most of the fetal life. D3 deiodinates T_3 and T_4 in the inner ring, leading to inactive compounds, and is a key enzyme during the fetal period. It is activated in proliferating states, and is present in high amounts in placenta, uterus and fetal membranes [6, 7]. The role of D3 is to act as a 'barrier', which prevents excessive amounts of maternal thyroid hormones from reaching the conceptus and keeps T_3 low throughout fetal development.

Thyroid hormones are important for normal development and the role of the deiodinases during development is to achieve the timely and tissue-specific appropriate T_3 levels. For example, D2 peaks in the mouse cochlea before the onset of hearing; D2 is thought to be involved in the maturation of the auditory function. In the rat brain, D2 activity peaks near birth at E21 and at day 15 after birth, dates that coincide with important periods of neuronal and glial maturation [4]. High D2 and T_3 concentrations are found in fetal BAT during the active recruitment of the tissue occurring before birth. Many other tissues such as the skin, the retina or the bone are also being studied. The role of D2 and D3 during fetal development is being studied in mice with targeted deletion of D2 and D3 [6]. The D2 knockout mice have pituitary resistance to T_4, impaired auditory and visual functions, become hypothermic in the cold and have reduced anxiety levels. The D3 knockout mice are hypothyroid, have growth retardation and impaired fertility associated with D3 being an imprinted gene.

In the present contribution we try to summarize what is known, and what is still unknown, regarding early and late fetal thyroid hormone physiology, its importance in neurodevelopment, its dependence on the production of T_4 by the mother, and the likely consequence of the untimely interruption caused by premature birth.

The Influence of Maternal Thyroid Hormones until Midgestation

Irreversible brain damage is found when thyroid hormone deficiency occurs during brain development. Epidemiological findings in areas of neurological cretinism clearly indicate an early involvement of the maternal thyroid hormones in fetal CNS development [2, 8, 9]. The severity of the CNS damage found is related to the degree and timing of maternal T_4 deficiency and is only prevented when the maternal hypothyroxinemia is corrected before midgestation [tables 1 and 2 in 2]. These ideas were difficult to reconcile with the success of the prompt postnatal treatment with T_4 of children with congenital hypothyroidism (CH). This success was considered proof that the developing fetal brain did not need thyroid hormone until after birth, because no major CNS damage was observed if the athyrotic newborn was promptly treated with T_4. However, new insights into maternal transfer of thyroid hormones throughout gestation and its likely role in fetal neurodevelopment have reconciled the findings in CH and in neurological cretinism caused by iodine deficiency.

Findings from Experimental Rat Models

Before onset of fetal thyroid function (FTF), T_4 and T_3 of maternal origin are present at very low concentrations in rat embryonic and fetal tissues, including the brain, that are directly influenced by the maternal serum T_4 [10]. The thyroid hormone receptor isoforms are already present long before the onset of FTF, at neural tube closure, and are likely to mediate biological effects of the T_3, locally generated from T_4 [11–13].

After onset of FTF, maternal transfer of thyroid hormones continues until term and represents an important proportion of thyroid hormones available to the fetus. In case of fetal thyroid failure, the amount of maternal T_4 reaching the fetal brain is enough to *selectively* prevent the cerebral T_3 deficiency of a hypothyroid rat [14]. Maternal T_4 and T_3, however, are not equivalent for the fetal brain, because during fetal and early postnatal development, cerebral T_3 depends on its local generation from T_4 – through D2 activity – and is not affected by circulating T_3 levels. Therefore, if the mother is hypothyroxinemic, the brain of a hypothyroid fetus is T_3-deficient, even if maternal and fetal circulating T_3 is normal or actually increased. Fetal brain T_3 is also protected from an excess of maternal T_4. Such results suggest that overtreatment of the mother with T_4 is less damaging for the fetal brain than maternal hypothyroxinemia. In contrast, there is almost no protection of the fetal brain from an excess of circulating T_3 [15].

If such experimental findings were relevant for man, they would explain why in most cases of promptly treated CH there is no permanent severe CNS damage. Most CH fetuses have a normal mother, supplying enough T_4 to the developing brain throughout gestation to avoid cerebral T_3 deficiency. As a result, the fetal brain has not been severely damaged before birth, and its normal development is thereafter achieved with T_4 treatment. These observations would also explain the irreversible damage caused by an insufficient supply of T_4 during early development, when the mother is the only source of hormone to the brain. Indeed, in the rat important phases of the development of the neocortex are altered by a period of maternal hypothyroxinemia preceding the onset of FTF [16, 17], showing directly that T_4 of maternal origin is important for early neurodevelopment. The most severe damage would be expected to occur when both the mother and fetus are hypothyroxinemic throughout pregnancy, as actually confirmed in humans [18].

Findings in the Human Fetus

During the last decades, our knowledge regarding thyroid hormone economy of the human fetus has increased both quantitatively and qualitatively and contributed to important new insights.

Major technical advances have made this possible, among them a) the development of highly sensitive methods to estimate very low concentrations of iodothyronines in fetal fluids and tissues (for which commercial kits are inadequate) and b) the development of transvaginal, ultrasound-guided puncture of the embryonic cavities to obtain samples from the fetal compartment without severing vascular connections with the mother [8, 9]. The latter option has changed some previously held concepts, which had been based on findings from aborted fetuses. We will artificially divide the information into two periods, namely the first half of gestation, when the mother is the major source of thyroid hormones available to the fetus, and the second half, when active FTF starts and the maternal contribution is still quite important. Most of the information is focused on the developing brain.

From Conception to Midgestation

Thyroxine, T_3 and reverse 3′,5′,3-triiodothyronine (rT_3) have been found in coelomic and amniotic fluids [19] from 5.8–11 weeks' postmenstrual age (PMA), which is 3.8–9 weeks' postconceptional age. The concentration of T_4 in the coelomic fluid is positively correlated with the concentration of the hormone in the mother's circulation, but values are extremely low compared with those of the mother [19]. The concentration of T_3 is at least 10-fold lower than that of T_4, whereas that of rT_3 is clearly higher. Concentrations in the coelomic fluid were higher than in the amniotic compartment. These low concentrations of T_4 and T_3 confirmed the efficiency of the placental 'barrier'. Because of the minute amounts of iodothyronines found in these fluids, their possible biological significance was often questioned, and therefore a second study included serum samples up to 17 weeks' PMA [19]. It confirmed that the concentration of T_4 in fetal fluids, including serum, is more than 100-fold lower than in maternal serum, and the concentration of T_3 is even lower.

Free T_4 (FT_4), however, was found to reach concentrations that are biologically active in their mothers (see fig. 2 in [9]): FT_4 in the fetal fluids is determined by the very low concentration of the T_4-binding proteins and the maternal T_4 that has escaped through the placental 'barrier'. The T_4-binding capacity of these proteins is determined ontogenically, is independent from maternal thyroid status, and is far in excess of the amounts of total T_4 that reach the fetal fluids. Thus, the availability of FT_4 to embryonic and fetal tissues is ultimately determined by the maternal circulating T_4 or FT_4, and will decrease in hypothyroxinemic women, even if clinically euthyroid [19]. The results obtained in these studies also explain why an efficient 'barrier' to maternal thyroid hormone transfer is actually necessary: without it the developing tissues would be exposed to inappropriately high, and possibly toxic, concentrations of free iodothyronines. Both a decrease and an inordinately high increase in the availability of FT_4 and/or FT_3 could result in adverse effects on the timely sequence of thyroid hormone-sensitive developmental events in the human brain.

It was also observed [19] that the concentration of TSH circulating in the fetal serum, obtained before severing maternal-fetal vascular connections, was very high, ranging from 2.9 to 7.2 mIU/l, and remained higher than in the maternal serum throughout pregnancy, confirming those values reported by Thorpe-Beeston et al. [20] using samples collected by cordocentesis, without disturbing maternal-fetal connections.

The presence of thyroid hormone receptors early in the development of the human fetal brain supports the hypothesis that thyroid hormone-sensitive developmental events might already occur before midgestation. These receptors were detected in the earliest samples of the cerebral cortex studied by Bernal et al. [11] at 9 weeks' PMA, with their concentrations increasing at least 10-fold by 18 weeks. The occupation of the thyroid hormone receptors by T_3 was 25–30% throughout the study period, despite the very low serum concentrations of this iodothyronine. This very important finding supports the hypothesis that the biological effects of the hormone might already occur in the cerebral cortex during the first trimester of human pregnancy. This possibility is supported by the more recent study by Iskaros et al. [13], which confirmed the early expression of TR genes in the whole fetal brain studied between 8.1 and 13.9 weeks PMA [11–13, 21].

The ontogenic patterns of the concentrations of T_4, T_3, rT_3, and the activity of the iodothyronine deiodinases D1, D2 and D3 have now been studied in nine different areas of the brain between 13 and 20 weeks' PMA. The developmental patterns of the iodothyronines, and of the activity of D2 and D3, showed both spatial and temporal specificity, but with divergence in the cerebral cortex and other brain areas, especially the cerebellum [22] (fig. 1). T_3 increased in the cerebral cortex between 13 and 20 weeks' PMA reaching concentrations similar to those in adult cortex, despite the very low concentration of circulating T_3. The data support the idea that T_3 in the human cerebral cortex is also locally generated from T_4, with considerable D2 activity being indeed found together with very low D3. In contrast, T_3 in cerebellum was very low, and increased only after midgestation, probably because cerebellar D3 activities were the highest of those found in the brain areas studied, and decreased only after midgestation. These findings support the hypothesis that T_3 is indeed required by the human cerebral cortex before midgestation, when the mother is the only source of the FT_4 available to fetal tissues, and confirm the important opposite roles of D2 and D3 in the local and timely bioavailability of cerebral T_3 during fetal life. It is important to realize that during the first trimester there is a maternal FT_4 peak [see the chapter by Glinoer, pp. 62–85] that appears imposed by the conceptus to ensure enough T_4 for generation of T_3 in the cerebral cortex up to midgestation [8, 9].

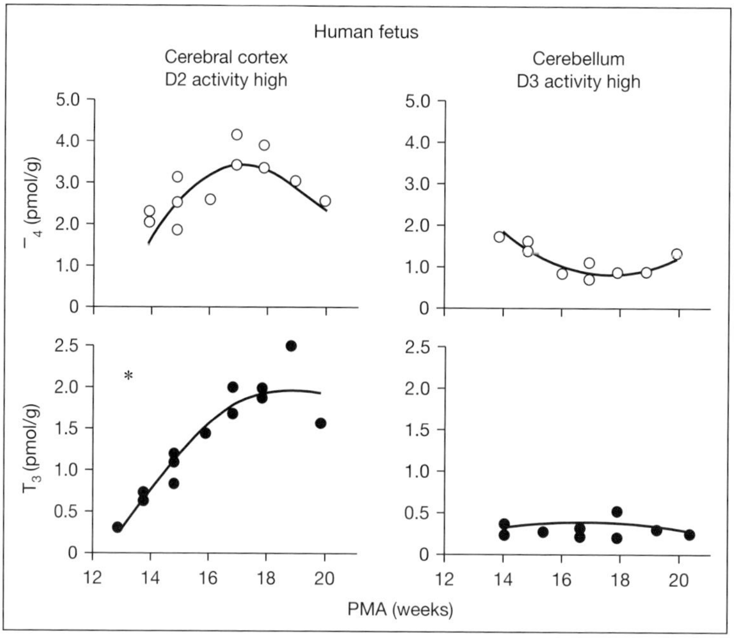

Fig. 1. Changes in the concentrations of T_4 and T_3 in the cerebral cortex and cerebellum of human fetuses before onset of FTF. T_4 in fetal serum increases 5-fold, from 3 to 15 pmol/ml; T_3 in fetal serum is very low throughout, approximately 0.5 pmol/ml (drawn using information from Kester et al. [22]). The asterisk indicates that the increase is significant.

Availability of Thyroid Hormones to the Fetus from Midgestation to Birth

Fluids of the Fetal Compartment and Fetal Brain

Most diagrams summarizing changes in circulating levels of T_4, T_3, rT_3, and TSH of the human fetus had for many years been based on serum samples from premature babies who had died for different reasons at variable intervals after birth and at different stages of gestation, the results being thus affected by many factors other than developmental age. As already indicated, ultrasound-guided blood sampling without interruption of the maternal to fetal vascular connections, finally assessed the fetal thyroid hormone situation in vivo [19, 20], confirming only some of the patterns previously described. Thorpe-Beeston et al. [20] confirmed that T_3 and FT_3 serum concentrations were very low throughout fetal life. In striking contrast to previous reports, however, T_4 and FT_4 were found to reach maternal and adult concentrations already at the beginning of the third trimester, and increased steadily with fetal age until birth (fig. 2). Fetal

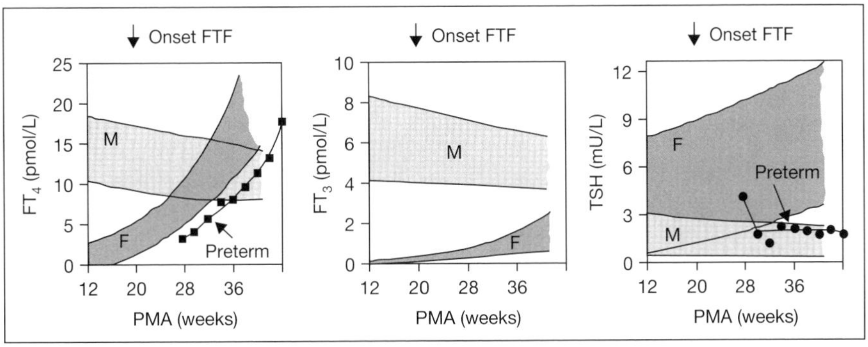

Fig. 2. Changes in the concentrations of FT_4, FT_3 and TSH throughout gestation, in maternal (M) and fetal (F) serum, before and after onset of FTF. The shaded areas enclose the values reported by Thorpe-Beeston et al. [20], obtained by cordocentesis without interruption of maternal-fetal vascular connections. Black squares and circles correspond to serum values found in premature infants, as reported by Morreale de Escobar and Ares [33].

serum TSH and FT_4 concentrations were found *positively* correlated ($r = 0.896$, $p < 0.001$) until birth and not negatively, as previously thought. Moreover, fetal serum TSH concentrations throughout pregnancy were well above the maternal TSH levels, reaching up to 12 mIU/l near term [20] (fig. 2).

Fetal T_4 and FT_4 are already increasing steadily in utero before the fetal thyroid itself is able to maintain such serum concentrations: the degree of iodination of thyroglobulin and its T_4 and T_3 contents are very low before 42 weeks' PMA [23]. The fetal contribution to its serum T_4 and FT_4 would be smaller, the earlier the gestational age. This may well be an important factor in the neonatal hypothyroxinemia of premature infants that will be discussed further on.

Due to obvious ethical constraints there is very little information regarding thyroid hormone concentrations and iodothyronine deiodinase expression and/or activities in different fetal tissues during the second half of pregnancy. Studies performed so far, including our own [22], have relied on autopsy material of premature babies affected by many factors other than development age. For these reasons, we will not review information obtained so far, as it is not possible to define the ontogenic developments per se, free from confounding factors.

The Role of the Mother from Midgestation to Birth

The transfer of maternal T_4 to the fetus continues until the umbilical cord is severed, as conclusively shown in 1989 by Vulsma et al. [24] who found concentrations of T_4 in cord blood of 7 neonates with complete organification defect that represented about 30–60% of the mean values reached by the normal fetus at term. In hypothyroid rat fetuses, serum T_4 concentrations that are

30–60% of those of normal fetuses, together with the response of D2 activity in the brain, are enough to preferentially avoid cerebral T_3 deficiency until birth [14]. Extrapolation of the latter findings to the human fetus suggests that after midgestation a normal maternal supply of T_4, together with the increase in cerebral D2 activity that occurs when the fetal thyroid does not secrete enough hormone, is sufficient to protect the brain from T_3 deficiency, and the accompanying CNS damage, until birth, explaining the good results of prompt postnatal treatment of CH infants.

Although valuable new insights have been obtained regarding the ontogenic patterns of cerebral thyroid hormone concentrations, their nuclear receptors, and the roles of the deiodinating isoenzymes in tailoring the bioavailability of T_3 to the developmental requirements of different cerebral structures, it is likely that we are still quite far from understanding all the mechanisms that may be involved, and their interrelationships. Very little is known regarding the role, in determining the availability of circulating T_4 to the fetal brain [for a review, see 9], of the activities of the deiodinating enzyme isoforms in other fetal tissues, as well as those of the sulfotransferases, glucoronidases, and sulfatases. Even less is known regarding the possible role of the recently identified specific iodothyronine plasma membrane transporters into, and out of, the fetal brain. We have already remarked upon our ignorance with respect to a possible developmental role of the high levels of TSH throughout gestation, as well as the cause for their rapid decrease after premature birth [9]. We still have insufficient information regarding the capacity of the fetal thyroid to meet the needs of the newborn preterm infant faced with the untimely interruption of the maternal supply of hormone, or how to improve it.

The Brain of the Fetus from Midgestation to Birth

In the rat, so often successfully used as experimental models for the study of the influence of thyroid hormones on brain development, the maturation of the brain is severely and irreversibly impaired when the postnatal thyroid function of the pup is inadequate. During the first few weeks of the suckling period brain development comprises phases that occur during the second half of gestation in humans. Figure 3a shows the human cerebral cortex at different stages of fetal development, specifically the development of layer V pyramidal neurons [25]. It clearly illustrates that major phases of corticogenesis still have to develop in the brain of the premature neonates as compared to those of the children born at term. Figure 3b and c also shows the complexity of changes between birth and weaning. Taking into consideration that an inadequate supply of thyroid hormones, especially of T_4, during the first half of human gestation, and that a delay in postnatal treatment with T_4 of congenitally hypothyroid newborns both result in central nervous system damage, it seems fair to conclude that neurodevelopment during the second half of pregnancy also requires an adequate supply of maternal

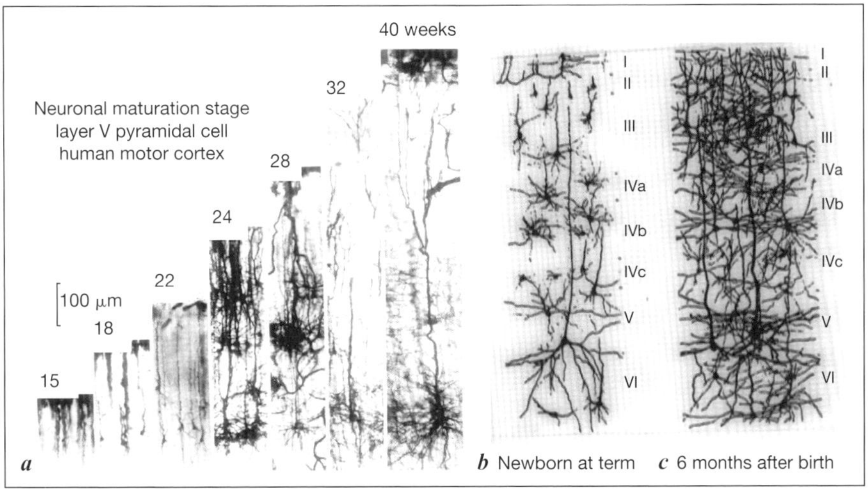

Fig. 3. ***a*** Maturation stage of a layer V pyramidal cell in the human motor cortex at different gestational ages [25]. It clearly illustrates the immaturity of these neurons in premature neonates born at 24–28 weeks of gestation. ***b*** The same type of pyramidal cell at birth. ***c*** At weaning (6 months after birth).

thyroid hormones. This conclusion receives direct experimental support in rats, where maternal hypothyroxinemia late in pregnancy, during a developmental period comparable to that occurring in the human brain during the second half of pregnancy, results in important irreversible alterations of cerebral cortex and hippocampal structures, and of behavior [16, 17].

Prematurity

Survival of premature infants has increased drastically with the improvements that have been introduced over the last decades, such as antepartum steroids, surfactant replacement, minimized volutrauma, optimized fluid maintenance, transfusion and nutritional intake, reduction in infections, and neonatal management techniques that emphasize stress reduction. This is also true for the survival of infants born at 23–27 completed weeks of gestation, with <1,000 g at birth, who we refer to here as 'great prematures'. But their improved survival rate has carried considerable costs [26], both emotional and economical, for the children themselves, their families and society: a large proportion of these children suffer from long-term disabilities, including disabling cerebral palsy; the only clearly associated factor is male sex. The results obtained in many other similar studies were also rather discouraging, and 'defining the limits of hope' [27].

Several observations, however, have pointed to the identification of causal factors that might be amenable to interventions ameliorating the developmental outcome of premature infants. During the last two decades several studies have been published that strongly support a causal connection between low circulating T_3 or T_4 [28, 29] during neonatal development of the premature infant and permanent cognitive and/or neurological abnormalities, including disabling cerebral palsy. The conclusions from these studies, was that the transient hypothyroxinemia or hypotriiodothyroninemia frequently accompanying prematurity should not be considered either 'physiological' or 'harmless'.

Although for years the low levels of circulating T_4 or T_3, without increased serum TSH, had been considered a 'physiological' consequence of fetal hypothalamic-pituitary-thyroid immaturity, results such as those illustrated in figure 2, however, strongly suggest an important role of the sudden interruption of the maternal supply of thyroid hormone, at phases of development when the thyroid hormone requirements of the neonate cannot be adequately met because of the neonate's hypothalamic-pituitary-thyroid immaturity. The 'physiological' situation of the premature infant, that is, to continue developing in utero, would have been quite different. Circulating T_4 and FT_4 are significantly higher for the fetus in utero than for age-matched prematurely born neonates. Moreover, the fetus is exposed to very high TSH levels, which drop abruptly with interruption of the maternal-fetal vascular connections. More recent studies performed on 620 premature infants [30] have confirmed that at 7 days after birth many of them had serum T_4 concentrations that are below those of term babies, and that 41% of neonates of the 23- to 27-week group had T_4 values below -1 SD of the cord levels adjusted for gestational age, TSH being also lower. Such results confirm that the intrauterine availability of both T_4 and T_3 is higher than that provided by the immature fetal thyroid, and that their postnatal hypothyroxinemia is not 'physiological', leading to a different outlook, in which the premature interruption of maternal transfer of thyroid hormones acquires an important causative role in the postnatal hypothyroxinemia of 'great' prematures.

It also became increasingly evident that this condition should not be considered 'harmless', because of the possibility that it is causally related to developmental deficits [28, 29]. The possibility that postnatal substitution therapy with thyroid hormones might ameliorate the outcome was explored. Among these studies, the randomized, placebo-controlled, double-blind trial of thyroxine supplementation in 200 infants born at less than 30 weeks' gestation, performed by van Wassenaer et al. [31] merit special attention. Half of the cohort was treated for 6 weeks after birth with T_4, the other 100 with placebo. Initial results at 24 months of age showed an important benefit of treatment in the few premature babies born at <27 weeks' gestation, with an 18 points higher developmental score, that reached normal values. In contrast, postnatal treatment of older prematures suggested negative effects. Developmental evaluation at 10 years of age confirmed the initial results, and supported the need for new double-blind treatment trials, especially in infants born at 27 weeks or less. A trial enabling project is at present being carried out to define doses and modes of postnatal treatments of such infants, in order to mimic the circulating levels of thyroid hormones that they would have if they were still developing in utero [32]. It is hoped that such attempts might change the present pessimistic 'definitions of the limits of hope' [27] for such infants.

References

1 Santisteban P: Development and anatomy of the hypothalamus-pituitary-thyroid axis; in Braverman LF, Utiger RD (eds): Werner and Ingbar's The Thyroid. A Fundamental and Clinical Text, ed 9. Philadelphia, Lippincott Williams & Wilkins, 2005, pp 8–25.

2 Morreale de Escobar G, Obregón MJ, Escobar del Rey F: Is neuropsychological development related to maternal hypothyroidism, or to maternal hypothyroxinemia? J Clin Endocrinol Metab 2000;85:3975–3987.

3 Ruiz de Ona C, Morreale de Escobar G, Calvo R, Escobar del Rey F, Obregon MJ: Thyroid hormones and 5′-deiodinase in the rat fetus late in gestation: effects of maternal hypothyroidism. Endocrinology 1991;128:422–432.

4 Obregon MJ, Ruiz de Ona C, Calvo R, Escobar del Rey F, Morreale de Escobar G: Outer ring iodothyronine deiodinases and thyroid hormone economy: responses to iodine deficiency in the rat fetus and neonate. Endocrinology 1991;129:2663–2673.

5 Ruiz de Oña C, Obregón MJ, Escobar del Rey F, Morreale de Escobar G: Developmental changes in rat brain 5′-deiodinase and thyroid hormones during the fetal period: the effects of fetal hypothyroidism and maternal thyroid hormones. Pediatr Res 1988;24:588–594.

6 Galton VA: The roles of the iodothyronine deiodinases in mammalian development. Thyroid 2005;15:823–834.

7 Huang SA, Dorfman DM, Genest DR, Salvatore D, Larsen PR: Type 3 iodothyronine deiodinase is highly expressed in the human uteroplacental unit and in fetal epithelium. J Clin Endocrinol Metab 2003;88:1384–1388.

8 Morreale de Escobar G, Obregón MJ, Escobar del Rey F: Maternal thyroid hormones early in pregnancy and fetal brain development. Best Practice Res Clin Endocrinol Metab 2004;18: 225–248.

9 Morreale de Escobar G, Obregon MJ, Escobar del Rey F: Role of thyroid hormone during early brain development. Eur J Endocrinol 2004;151(suppl 3):U25–U37.

10 Morreale de Escobar G, Pastor R, Obregón MJ, Escobar del Rey F: Effects of maternal hypothyroidism on the weight and thyroid hormone content of rat embryonic tissues. Endocrinology 1985;117:1890–1901.

11 Bernal J, Pekonen F: Ontogenesis of the nuclear 3,5,3′-triiodothyroxine receptor in the human fetal brain. Endocrinology 1984;114:677–679.

12 Bernal J, Pérez-Castillo A, Pans T, Pekonen F: Ontogenesis of thyroid hormone receptor; in Labrie F, Proulx L (eds): Endocrinology. Amsterdam, Elsevier Science Publisher, 1984, pp 977–980.

13 Iskaros J, Pickard M, Evans I, Sinha A, Hardiman P, Ekins R: Thyroid hormone receptor gene expression in first trimester human fetal brain. Journal of Clinical Endocrinology and Metabolism 2000;85:2620–2623.

14 Calvo R, Obregon MJ, Ruiz de Oña C, Escobar del Rey F, Morreale de Escobar G: Congenital hypothyroidism, as studied in rats. Crucial role of maternal thyroxine but not of 3,5,3′-triiodothyronine in the protection of the fetal brain. J Clin Invest 1990;86:889–899.

15 Morreale de Escobar G, Calvo R, Obregon MJ, Escobar del Rey F: Homeostasis of brain T3 in rat fetuses and their mothers: effects of thyroid status and iodine deficiency. Acta Med Austriaca 1992;1:110–116.

16 Lavado-Autric R, Ausó E, García-Velasco JV, Arufe MC, Escobar del Rey F, Berbel P, Morreale de Escobar G: Early maternal hypothyroxinemia alters histogenesis and cerebral cortex cytoarchitecture of the progeny. J Clin Invest 2003;111:1073–1082.
17 Auso E, Lavado-Autric R, Cuevas E, Del Rey FE, Morreale De Escobar G, Berbel P: A moderate and transient deficiency of maternal thyroid function at the beginning of fetal neocorticogenesis alters neuronal migration. Endocrinology 2004;145:4037–4047.
18 Yasuda T, Ohnishi H, Wataki K, Minagava M, Minamitani K, Niimi H: Outcome of a baby born from a mother with acquired juvenila hypothyroidism having undetectable thyroid hormone concentrations. J Clin Endocrinol Metab 1999;84:2630–2632.
19 Calvo RM, Jauniaux E, Gulbis B, Asunción M, Gervy C, Contempré B, Morreale de Escobar G: Fetal tissues are exposed to biologically relevant free thyroxine concentrations during early phases of development. Possible consequences of maternal hypothyroxinemia. J Clin Endocrinol Metab 2002;87:1768–1777.
20 Thorpe-Beeston JG, Nicolaides KH, Felton CV, Butler J, McGregor AM: Maturation of the secretion of thyroid hormone and thyroid-stimulating hormone in the fetus. N Engl J Med 1991;324:532–536.
21 Ferreiro B, Bernal J, Goodyer CG, Branchard CL: Estimation of nuclear thyroid hormone receptor saturation in human fetal brain and lung during early gestation. J Clin Endocrinol Metab 1988;67:853–856.
22 Kester MH, Martinez de Mena R, Obregon MJ, Marinkovic D, Howatson A, Visser TJ, Hume R, Morreale de Escobar G: Iodothyronine levels in the human developing brain: major regulatory roles of iodothyronine deiodinases in different areas. J Clin Endocrinol Metab 2004;89:3117–3128.
23 van den Hove MF, Beckers C, Devlieger H, de Zegher F, De Nayer P: Hormone synthesis and storage in the thyroid of human preterm and term newborns: effect of thyroxine treatment. Biochimie 1999;81:563–570.
24 Vulsma T, Gons MH, de Vijlder JJM: Maternal-fetal transfer of thyroxine in congenital hypothyroidism due to a total organification defect or thyroid agenesis. N Engl J Med 1989;321:13–16.
25 Marin-Padilla M: Dual origin of the mammalian neocortex and evolution of the cortical plate. Anat Embryol (Berl) 1978;152:109–126.
26 Wood NS, Marlow N, Costeloe K, Gibson AT, Wilkinson AR: Neurologic and developmental disability after extremely preterm birth. EPICure Study Group. N Engl J Med 2000;343:378–384.
27 Cole FS: Extremely preterm birth – defining the limits of hope. N Engl J Med 2000;343:429–430.
28 Den Ouden AL, Kok JH, Verkerk PH, Brand R, Verloove-Vanhorick SP: The relation between neonatal thyroxine levels and neurodevelopmental outcome at age 5 and 9 years in a national cohort of very preterm and/or very low birth weight infants. Pediatr Res 1996;39:142–145.
29 Reuss ML, Paneth N, Pinto-Martin JA, Lorenz JM, Susser M: The relation of transient hypothyroxinemia in preterm infants to neurologic development at two years of age. N Engl J Med 1996;334:821–827.
30 Williams FL, Simpson J, Delahunty C, Ogston SA, Bongers-Schokking JJ, Murphy N, van Toor H, Wu SY, Visser TJ, Hume R: Developmental trends in cord and postpartum serum thyroid hormones in preterm infants. J Clin Endocrinol Metab 2004;89:5314–5320.
31 van Wassenaer AG, Westera J, Houtzager BA, Kok JH: Ten-year follow-up of children born at <30 weeks' gestational age supplemented with thyroxine in the neonatal period in a randomized, controlled trial. Pediatrics 2005;116:e613–e618.
32 La Gamma E, van Wassenaer A, Golombek SG, Morreale de Escobar G, Kok J, Quero J, Ares S, Paneth N, Fisher DA: Neonatal supplementation for transient hypothyroxinemia of prematurity (THOP): beneficial or detrimental? Treat Endocrinol 2006;5:335–346.
33 Morreale de Escobar G, Ares S: The hypothyroxinemia of prematurity. J Clin Endocrinol Metab 1998;83:713–716.

Dr. Maria Jesus Obregon
Instituto de Investigaciones Biomedicas
Centro mixto 'Alberto Sols' (CSIC-UAM)
Arturo Duperier, 4, ES–28029 Madrid (Spain)
Tel. +34 91 585 4449, Fax +34 91 585 4401, E-Mail mjobregon@iib.uam.es

Van Vliet G, Polak M (eds): Thyroid Gland Development and Function.
Endocr Dev. Basel, Karger, 2007, vol 10, pp 99–117

New Phenotypes in Thyroid Dyshormonogenesis: Hypothyroidism due to DUOX2 Mutations

José C. Moreno, Theo J. Visser

Department of Internal Medicine, Erasmus Medical Center, Erasmus University, Rotterdam, The Netherlands

Abstract

Hydrogen peroxide (H_2O_2) is an essential compound for the synthesis of thyroid hormone. Its presence in the follicular lumen is required by thyroperoxidase for the iodination of tyrosil residues of thyroglobulin, the initial step in the synthesis of T_3 and T_4. The biochemical requirement of H_2O_2 for thyroid hormone production has been known for decades and an H_2O_2-generating system was predicted to exist in the thyroid gland. In recent years, different research groups have unraveled the molecular nature of the system. Two homologous proteins, the dual oxidases 1 and 2, DUOX1 and DUOX2 (formerly THOX1 and 2, for thyroid oxidases), were identified and shown to contain functional domains typical of NADPH oxidoreductases. However, in vitro reconstitution of H_2O_2 production could not be obtained in nonthyroidal cell lines expressing these proteins. Evidence of DUOX involvement in thyroidal H_2O_2 production came from the identification of a DUOX2 *nonsense* homozygote mutation, which resulted in the absence of all functional domains of the protein, in a patient with permanent and severe congenital hypothyroidism (CH). Recently, an experimental demonstration of H_2O_2 production by DUOX2 was achieved by coexpression with a novel 'maturation factor' for DUOX2 (DUOXA2). Transient CH has also been linked to heterozygote *nonsense* DUOX2 mutations, showing for the first time that transitory CH can have a genetic origin. These findings also establish that partial dyshormonogenetic defects can behave biochemically as transient forms of CH. Novel *missense* and *splice-site* DUOX2 mutations in compound heterozygosity have been recently reported in association with a wide spectrum of hypothyroid phenotypes, ranging from very mild to severe. Functional analysis of these point mutations using available assays opens now the possibility to ascertain whether transiency or permanency of DUOX2 phenotypes relate exclusively to monoallelic or biallelic inactivation of the gene, or if the degree of pathogenic severity of mutations may also influence the timely outcome of this type of hypothyroidism.

Introduction

Congenital hypothyroidism (CH) is the most common congenital endocrine disorder, affecting 1 in every 3,000 newborns [1]. Detection by neonatal CH screening programs was widely implemented in western countries mostly from the 1980s, thus allowing its early detection and treatment, and preventing major irreversible cognitive and motor damage derived from the lack of thyroid hormone during early postnatal life [2].

Based on outcome and evolution of the disease, CH can be classified as permanent or transient. While the previous figure (1:3,000 neonates) represents the prevalence of permanent cases of CH and has been found to be very similar among countries, the prevalence of transient CH (TCH) notably differs between geographical areas of the world, ranging from approximately 1:25,000 to 1:2,250 neonates (table 1). These discrepancies in TCH prevalence among countries are attributed to the variable presence of iodine deficiency in the population, the use of iodine-containing compounds, the differences in CH screening protocols or the incomplete recording of transient cases of hypothyroidism [3–6]. Considering both permanent and transient forms of the disease, CH would reach a prevalence of 1:1,200 neonates in a country like the Netherlands, without reported iodine insufficiency (table 1) [7].

Permanent CH is mainly caused by alterations in the embryonic development of the thyroid gland (dysgenesis), but a substantial 20% of cases are due to biochemical defects in the process of thyroid hormone synthesis (dyshormonogenesis). The molecular etiology of thyroid dysgenesis is largely unknown, with only a few transcription factors and the TSH receptor having been involved in human cases of agenesis or hypoplasia of the gland [8]. On the other hand, the intricate network of cellular pathways that are active through the development of a mature thyroid gland is becoming increasingly understood from research involving manipulation of the mouse genome [8, 9]. As a consequence, the molecular etiology of thyroid dyshormonogenesis has become better understood in the last decades, with the identification of the proteins responsible for most essential biochemical steps in thyroid hormonogenesis [9].

The etiology of TCH is reportedly very variable [10]. TCH is common in prematurely born infants. Furthermore, it can be caused by maternal iodine deficiency, exposure to excessive amounts of iodine around the time of birth (e.g. by the use of iodinated compounds) and fetal exposure to maternally derived thyroid-blocking antibodies or antithyroid drugs in case of women with autoimmune thyroid disease. Rarely, protein-losing nephrosis can also induce TCH. However, for a substantial proportion of patients studied in large series of TCH around the world, the underlying etiology remained unknown (table 1).

Table 1. Estimated prevalence and etiology of transient congenital hypothyroidism (TCH) in large CH series

	USA/Canada	Australia	Netherlands	France
Study period	1972–1978	1977–1986	1981–1982	1982–1987
Method	T_4-TSH/TSH	T_4-TSH	T_4-TSH	TSH
Neonates screened	1,046,362	570,000	346,355	202,930
Prevalence	1:26,159	1:13,750	1:2,249	1:2,742
Etiology, %				
Iodine excess	57.5[a]	58.5	57	59
MH, ATD, TBA		8.5	3.5	4
Prematurity, LBW	–	excluded[b]	23	16
T_4 losses	–	–	2.5	–
Thyroid dysgenesis	22.5	4	–	–
'Idiopathic'	20	29	14	21
Reference	Fisher et al. [3]	Coackley et al. [4]	Vulsma [5]	Léger and Czernichow [6]

Of note, in a substantial percentage of the TCH population a known cause of TCH could not be traced. Thyroid dyshormonogenesis is not among the known causes of TCH. MH = Maternal hypothyroidism; ATD = autoimmune thyroid disease; TBA = maternal thyroid-blocking antibodies; LBW = low birth weight; – = not present.

[a]This includes iodine excess as well as MH, ATD and TBA.

[b]Premature babies were excluded from the cohort studied.

In the past, sporadic reports suggested that certain cases of TCH might be related to thyroid dyshormonogenesis [11], but this was never formally demonstrated. This chapter will briefly review the genetic basis of 'classical' phenotypes of thyroid dyshormonogenesis, but will focus on currently available knowledge over the thyroidal H_2O_2-generating system of dual oxidases (DUOX), genetic defects of which have been linked to cases of either permanent or TCH.

Overview of Molecular Thyroid Hormonosynthesis

The proteins responsible for most critical biochemical steps in the synthesis of thyroid hormones have been identified (fig. 1). Our mechanistic understanding of the function of these proteins has increased considerably. To the long-established roles of thyroglobulin (Tg) and thyroperoxidase (TPO) as respective structural and enzymatic pillars for T_4 and T_3 production, novel roles for newly identified proteins emerged in the last decade. Iodide transporters located at the basal and apical membranes of the thyrocyte, named NIS and

a

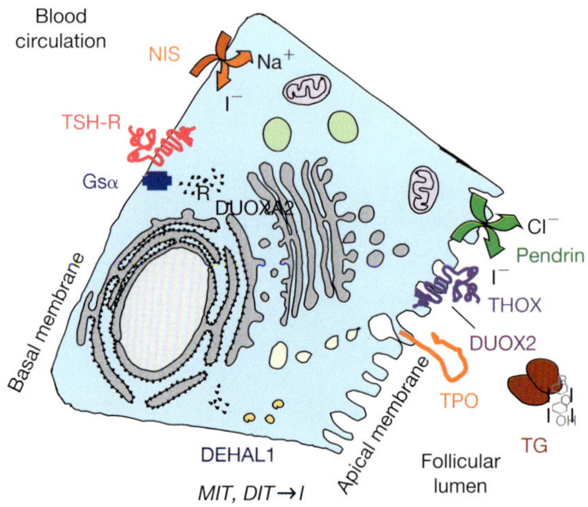

b

Gene	Locus	Protein function	Phenotype	Inher.
TSH-R	14q31	Activation of various thyroid-specific metabolic pathways	– Euthyroid hyperthyrotropinemia – Thyroid hypoplasia	A.R. A.D. A.R.
GNAS1	20q13	*Gs proteins*: Signal transduction from GPCRs for stimulation of adenylyl cyclase	Resistance to TSH and/or *Albright hereditary osteodystrophy*	A.D.
NIS	19p13	Basal transport of iodide from bloodstream into the thyroid cell	– Severe or moderate CH – Euthyroid goiter	A.R.
TG	8q24	Structural support (pro-hormone) for thyroid hormone synthesis	– Goiter and severe or moderate CH – Euthyroid goiter; occasionally PIOD	A.R. A.D.
TPO	2p25	Iodination of tyrosine residues of thyroglobulin (iodide organification)	– Severe CH – Total iodide organification defect (TIOD)	A.R.
PDS	7q31	Apical transport of iodide from the cytoplasm to the follicular lumen	– '*Pendred syndrome*': goiter and/or moderate hypothyroidism and deafness – Partial iodide organification defect (PIOD)	A.R.
DUOX2	15q15	Generation of H_2O_2 in the thyroid follicle	– Permanent and severe CH (TIOD) – Transient and moderate CH (PIOD) – Permanent and mild CH (PIOD)	A.R. A.D. A.R.
DUOXA2	15q15	Transition from ER to Golgi, maturation and membrane localization of DUOX2	Not described	–
DEHAL1	6q24	Deiodination of iodotyrosines (MIT, DIT) for intracellular recycling of iodide	Iodotyrosine dehalogenation deficiency	A.D.

pendrin, respectively, have provided the molecular basis for the dynamics of iodide transfer into the cytoplasm of the thyroid cell and towards the follicular lumen. The pivotal roles of the TSH receptor and its coupled Gs proteins in the postnatal proliferation of thyroid cells and in the activation of signal pathways specific to thyroid metabolism were also unraveled. Recently, the identification of two oxidase proteins, originally named THOX1 and THOX2, which are involved in H_2O_2 generation at the apical membrane of the thyrocyte, identified the molecular nature of a biochemical activity that had been known to exist for decades. Finally, the molecular basis for the 'dehalogenation of iodotyrosines', an activity that recycles iodide within the thyroid gland through the deiodination of mono- and di-iodotyrosines, has been recently established with the cloning of a gene named DEHAL1 [12].

Functional inactivation of most of these genes has been identified in patients with CH (fig. 1b). Each defect leads to a particular type of CH, which can be accompanied by features in different organs, such as deafness in Pendred's syndrome. Inheritance of dyshormonogenic defects usually follows a mendelian recessive pattern, with some rare exceptions (fig. 1b). Prior to the identification of clinical phenotypes derived from defects in H_2O_2 generation, all CH phenotypes due to thyroid dyshormonogenesis were described as permanent.

Identification, Structural Features and Function of Dual Oxidases (DUOX1 and DUOX2)

Since H_2O_2 is a reactive oxygen species known to be necessary for thyroid hormone synthesis, thyroid-related research groups originally described the cloning of DUOXes. Dupuy et al. [13] first reported the cloning of the pig and human $p138^{Tox}$ flavoproteins through protein purification, microsequencing and rapid amplification of cDNA ends from thyroid tissue. De Deken et al. [14] thereafter described the cloning of two homologous human cDNAs, named THOX1 and THOX2 (for thyroid oxidases), through the screening of a human thyroid cDNA library with a probe specific for $gp91^{phox}$, a well-studied oxidase

Fig. 1. Molecular overview of thyroid hormonogenesis. ***a*** Proteins involved in thyroid hormone synthesis. Most relevant steps in thyroid hormonogenesis have been identified at the molecular level. ***b*** Gene localization, clinical phenotypes of CH and mode of inheritance (Inher.) linked to genetic defects in humans. Clinical phenotypes presented here are only caused by *inactivating* gene mutations; phenotypes associated with hyperstimulation or overexpression by activating mutations are not shown (modified with permission of Elsevier [9]). GPCR = G protein-coupled receptor; MIT/DIT = mono-/di-iodotyrosine; A.R. = autosomal recessive; A.D. = autosomal dominant.

a

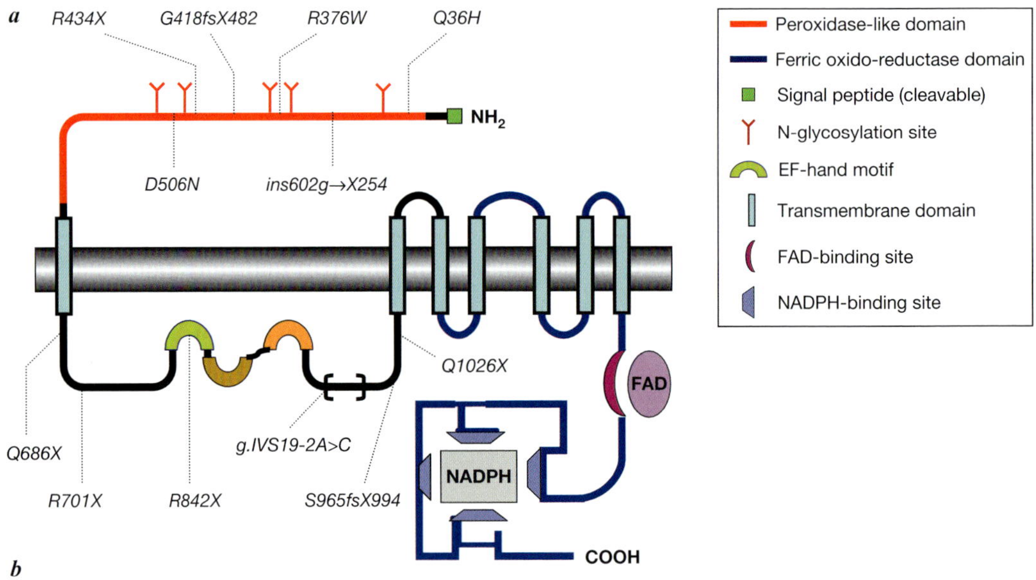

b

DUOX2 mutation	Mutation state	Perchlorate discharge	CH phenotype	Reference
R434X	Homozygous	100% (TIOD)	Permanent severe	Moreno et al. [10]
Q686X	Heterozygous	66% (PIOD)	Transient mild	Moreno et al. [10]
R701X	Heterozygous	41% (PIOD)	Transient mild	Moreno et al. [10]
S965fsX994	Heterozygous	40% (PIOD)	Transient mild	Moreno et al. [10]
ins602g→X254	Heterozygous	n.d.	To be determined*	Pfarr et al. [29]
S965fsX994	Heterozygous	20–63% (PIOD)[+]	To be determined*	Di Candia et al. [30]
Q1026X	Heterozygous	20-63% (PIOD)[+]	To be determined*	Di Candia et al. [30]
R842X + R376W	Comp. heterozygous	28% (PIOD)	Permanent mild	Vigone et al. [27]
S965fsX994 + Q36H	Comp. heterozygous	46% (PIOD)	Permanent mild	Varela et al. [28]
G418fsX482 + g.IVS19-2A>C	Comp. heterozygous	68% (PIOD)	Permanent	Varela et al. [28]
ins602g→X254 + D506N	Comp. heterozygous	n.d.	To be determined[#]	Pfarr et al. [29]

in phagocytes, active in host defense. THOX2 turned out to be the full-length version of the partial p138^{Tox} sequence. Applying serial analysis of gene expression (SAGE) to human thyroid tissue, subtraction of tissue-specific SAGE tags (individual mRNAs overrepresented in thyroid) and screening of a thyroid cDNA library with SAGE tag-related ESTs (expressed sequence tags), a clone containing THOX2 3′-sequences was obtained, showing preferential expression of the human gene in thyroid tissue [15].

Human THOX genes are located on chromosome 15q15.3, only 16 kb apart from each other and in opposite transcriptional orientations. THOX1 comprises around 36 kb and contains 35 exons (of which the first two are noncoding), while THOX2 spans over 22 kb and has 34 exons (the first one being noncoding). THOX1 and THOX2 proteins contain 1,551 and 1,548 amino acids, respectively, showing 84% sequence similarity.

Edens et al. [16], who cloned the homologous sequences in *Caenorhabditis elegans*, suggested a change of nomenclature (DUOX, for dual oxidases) based on the structural features of these proteins, since two distinct parts can be distinguished in DUOXes: a C-terminal segment with high homology to NADPH oxidases and an N-terminal segment homologous to peroxidases (fig. 2a). Supporting this change in nomenclature, expression of DUOXes was shown not to be restricted to the thyroid [15, 17]. Immunohistochemistry localized DUOX proteins in the apical membrane of the thyrocyte [14] with antibodies cross-reacting with the 2 proteins. Based upon the proposed topographic model, the amino-terminal peroxidase domain would be extracellular, facing the follicular lumen, and the carboxy-terminal tail of the oxidase domain, intracellular (fig. 2a).

DUOXes are glycoproteins with seven putative transmembrane domains, between the first and second of which a large intracellular loop exists containing EF-hand motifs for calcium binding. The C-terminal, intracellular segment contains typical oxidase domains: 1 FAD-binding site and 4 NADPH-binding sites (fig. 2a). The presence of EF-hand motifs suggests that calcium ions

Fig. 2. Structural model for DUOX2 protein and mutations found in CH patients. ***a*** Structural and functional domains of DUOX2. Location of genetic defects found in CH patients at the amino acid level is indicated by dotted lines. The squared legend details the symbolic figures assigned to each motif or functional feature in DUOX2. The 2007 release of UniProtkB/Swiss-Prot for the functional topography of Q9NRD8 protein (DUOX2_HUMAN) recognizes 3 EF-hand motifs in DUOX2, only the first two of which have potential for Ca^{2+} binding. The g.IVS19-2A>C mutation is represented here as (putatively) skipping exons 20 and 21 of DUOX2, based on exon-trapping experiments (minigene). ***b*** Characteristics of known human DUOX2 mutations and essentials of human hypothyroid phenotypes associated with them. n.d. = Not determined; $^{+}$ = range of discharge in a group of patients; * = expected transient upon the genotype; $^{\#}$ = expected permanent upon the genotype.

regulate the enzyme activity. In dog thyrocytes, but not so much in the human, increasing cAMP levels by forskolin stimulate the expression of DUOXes [14].

Expression of either or of both DUOX proteins is currently documented in more than 20 tissues [17], but a distinct expression profile for each of the DUOXes among tissues is emerging that suggests differentially regulated expression. DUOX2 is highly expressed in thyroid but also along the intestinal tract, salivary glands and pancreas, while DUOX1 shows preferential expression over DUOX2 in airway epithelial tract and skin. Even when some tissues coexpress both DUOX (e.g. thyroid and epithelial airways), their pattern of expression strongly suggests 2 functionally independent proteins and separate physiological functions for DUOX1 and DUOX2 [17].

Investigations of the function of DUOXes has been hampered by the lack of a working cellular model where these proteins could be functionally expressed in heterologous cell lines. When DUOX enzymes were expressed in mammalian nonthyroidal cell systems, they remained in the endoplasmic reticulum (ER) and did not reach the orthotopic cellular location at the membrane [18]. Whereas the postulated function of DUOXes in the thyroid was the generation of H_2O_2 for thyroid hormone synthesis, this could not be experimentally confirmed by reconstitution of hydrogen peroxide in vitro. Recently, Grasberger and Refetoff [19] identified an ER-resident factor, named DUOXA2, which allows the transition of DUOX2 from ER to Golgi, its maturation and its final translocation to the plasma membrane. By coexpression of this new factor with DUOX2 in mammalian HeLa cells, they could reconstitute in vitro H_2O_2 production. A protein paralog (DUOXA1) was also identified as the maturation factor for DUOX1. Interestingly, genomic location of both DUOXA genes lies within the 16 kb that separates DUOX1 from DUOX2 on chromosome 15, and the ancient genomic rearrangement that evolutionarily linked DUOX and DUOXA genes suggests their coupled activity as a sort of eukaryotic operon-like functional unit.

Availability of this functional assay for DUOX2 has straightforward implications. First, it has now experimentally confirmed the involvement of DUOX2 in H_2O_2 generation (so far exclusively evidenced by finding DUOX2 defects in severely hypothyroid patients), but also shows that DUOX2, independently from DUOX1, suffices for the generation of hydrogen peroxide. Furthermore, it opens the opportunity for structure-function analyses on DUOX2, including the study of *missense* mutations and sequence variations with less obvious pathogenic effects than the early truncating stop-codon (*nonsense*) mutations originally described [10, 20] (fig. 2a).

While the initial expectation was that DUOXes would work in close association with each other and/or with other (putatively cytosolic) proteins in a coordinate system (as occurs with the gp91^{phox} complex in phagocytes [21]), experimental evidence now supports that membrane-located DUOX2 alone suffices for the generation of hydrogen peroxide. Thus, it seems now more doubtful that other cytosolic/membranal adjuvant proteins might cooperate with DUOX2 or be necessary components of the functional H_2O_2-generating machinery in the thyroid gland. This does not exclude physical interactions between DUOX2 and proteins like the thioredoxin-like EFP1 (identified by 2-hybrid screening using DUOXes as bait [22]) or the maturation factor DUOXA2 (which stays in the ER while DUOX2 proceeds towards insertion in the membrane [19]), which would only represent physical interaction with chaperon-like factors with the purpose of protein processing, but not a functional cooperation for H_2O_2 production.

Both DUOX isoforms are present in the thyroid gland, but we currently do not understand the functional significance of the presence of these two highly homologous proteins in the same tissue. No functional or structural differences have been described between DUOX1 and DUOX2 but, again, genotype-phenotype studies in CH patients suggest that DUOX1 is not involved in H_2O_2 generation in the thyroid, since intact DUOX1 cannot functionally rescue the lack of activity of DUOX2 in patients with either severe or milder hypothyroidism [10].

In the synthesis of thyroid hormone, TPO activity has been classically regarded as responsible for both the oxidation of iodine prior to iodination of tyrosil residues of Tg (iodide organification) and the linking of mono- or bi-iodinated tyrosines to form T_3 or T_4 (coupling). However, the interesting findings of Edens et al. [16] that DUOX homologues of *C. elegans* catalyze the crosslink of tyrosines on the cuticular extracellular matrix of the worm, inducing its stabilization, suggest the straightforward possibility that DUOXes could be involved, alone or in cooperation with TPO, in the coupling of iodotyrosines in the Tg molecule. In this respect, there is controversy as to whether the N-terminal peroxidase domain of DUOX can be active or not, based on the absence in DUOX sequences of highly conserved histidine residues that are necessary for heme binding in peroxidases [23]. However, recent evidence that the peroxidase-like domain of DUOX is active when expressed in bacteria [16] and also in eukaryotes (human respiratory tract epithelial cells [24]) further substantiates this hypothesis. Again, the in vitro functional study of natural or artificial point mutations in the peroxidase-like domain of DUOX2 (leaving the oxidase domain intact) might be pivotal in the analysis of this question.

DUOX2 Phenotypes of CH

To validate the aforementioned findings and investigate the clinical implications of novel DUOXes, 9 patients diagnosed with CH by neonatal screening were selected for mutational analysis of DUOX1 and DUOX2 genes [10]. All of them showed a thyroid gland in situ and all had a positive perchlorate discharge test, indicating that their dyshormonogenesis was due to an iodide organification defect, either total (TIOD, 100% discharge: in 1 patient) or partial (PIOD, 20–90% discharge, in 8 patients). These patients did not show clinical features, biochemical signs or molecular findings suggesting defects in TPO, Tg or PDS genes.

The patient with TIOD had a severe form of CH and harbored a homozygous *nonsense* mutation in DUOX2 (R434X) which prematurely truncates the protein before coding of the first transmembrane domain (fig. 2a). His clinical phenotype is indistinguishable from the classical phenotype of homozygote TPO defects. This biallelic mutation encoding a prematurely DUOX2 termination signal is the only in vivo evidence in humans establishing the role of DUOX2 in thyroid hormonogenesis. The absence of DUOX2 activity in this patient, which cannot be compensated by an intact DUOX1, is shown to completely block the synthesis of thyroid hormone. The exclusive thyroidal phenotype in this girl suggests that presence of DUOX2 in other tissues, like digestive tract or pancreas, is not essential or can at least be compensated by other sources of reactive oxygen species in these tissues.

Three CH patients with PIOD were found to have heterozygous *nonsense* or *frameshift* mutations (Q686X, R701X and S965fsX994) also leading to premature truncation of most functional domains of the NADPH oxidase segment of DUOX2 (fig. 2a). Interestingly, these 3 patients (as well as the additional 5 studied) showed a transient form of CH [10]. Their blood spot T_4 and TSH levels in the Dutch screening program showed a milder type of CH, confirmed later through plasma thyroid function tests. A high uptake of ^{123}I and washout of 40–65% indicated dyshormonogenesis due to PIOD. These three baby girls were started on T_4 substitution treatment, but T_4 dosage requirements became low (1.2–1.4 μg/kg/day) compared to most CH cases. At ages 3–5 years, T_4 withdrawal trials showed they could all maintain euthyroidism for 2 months without thyroxine. These findings for the first time showed that TCH can be a genetic disease, and disclose that at least a proportion of the etiologically 'idiopathic' cases of TCH (table 1) correspond to monoallelic DUOX2 defects causing thyroid dyshormonogenesis.

The apparent paradox that an obviously permanent genetic defect could cause a transient phenotype of disease deserved justification. We believe that the most likely explanation relates to the existence of well-documented, high requirements of thyroid hormone at the beginning of postnatal life (12 μg/kg/day), which progressively decrease during the first 6 months (6 μg/kg/day) to finally reach a plateau after the first year (at around 3 μg/kg/day) [25] (fig. 3). In this setting, it is plausible to think that DUOX2 haploinsufficiency, reducing to 50%

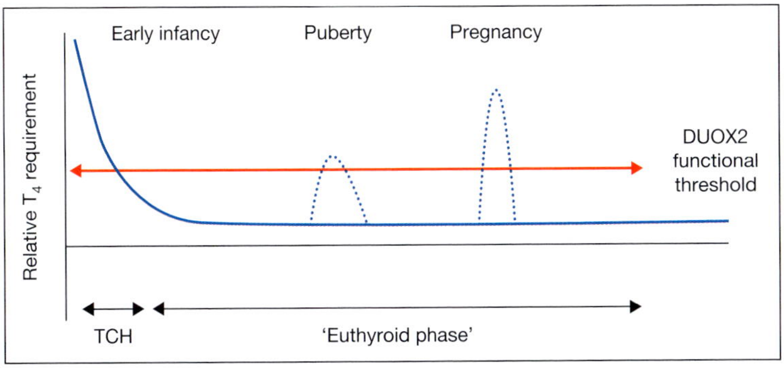

Fig. 3. Phenotypical features and possible outcome of hypothyroidism due to haploinsufficiency of DUOX2. In blue, high requirements for thyroid hormone at the beginning of life decrease dramatically during the first year of age to reach a plateau, but relative T_4 needs could reincrease during accelerated growth in puberty and during pregnancy. Depending on individual thresholds for H_2O_2 production (and T_4, T_3 synthesis), in red, neonatal-infantile TCH could be followed by a 'euthyroid' phase of the disease, which could be interrupted in circumstances such as increased thyroid hormone requirement, reduced iodine availability, thyroid gland damage by inflammation and others. This speculative scheme might not be exclusive for monoallelic DUOX2 defects, but could be also applicable to biallelic defects with diminished but residual activity in each allele.

the capacity to generate H_2O_2 and thyroid hormone, could be detectable at neonatal screening, but the severity of hypothyroidism might progressively diminish and gradually allow a sufficient H_2O_2 production which would herald the beginning of the euthyroid phase of the phenotype (fig. 3). However, a decreased capacity for hydrogen peroxide and T_3/T_4 production will characterize these patients throughout their lives. This, in combination with the fact that thyroid hormone requirements increase during pregnancy and maybe also during puberty (fig. 3), supports the notion that patients with monoallelic DUOX2 mutations need to be followed up and closely monitored for possible development of 'subclinical' or overt hypothyroidism after they have reached the euthyroid phase of the disease.

Timely detection of possible TSH elevations could prevent goitrogenesis but, most important, close monitoring of thyroid status of these patients during pregnancy might detect the development hypothyroxinemia, which would affect the psychomotor development of the fetus [26]. Furthermore, the fetus might also harbor the DUOX2 defect, narrowing its own chances for compensation of thyroid hormone shortage during the 2nd and 3rd trimesters of gestation. Speculation of a 'transient-recurrent' phenotype of hypothyroidism in DUOX2 defects might have to be proven in the future through long-term follow-up of these (and other) reported patients with TCH. At this point, thyroid function monitoring of pregnant mothers of DUOX2 patients with TCH who themselves carry the DUOX2 defect is clinically advisable and can shed light on the possibility of recurrence of this type of hypothyroidism in adulthood.

Novel DUOX2 Variants, Increased Variability of Phenotypes

Recently, eight novel mutations and one that has already been described in DUOX2 have been identified in 7 family pedigrees from different countries (fig. 2) [27–30]. For the first time, some mutations are reported in compound heterozygosity, while others are simple heterozygous. They comprise *nonsense* (2), *frameshift* (3) and missense (3) mutations, as well as one defect putatively affecting exon splicing. The already described heterozygous S965fsX994 frameshift (due to insertion of 4 base pairs in exon 21 of DUOX2) is now present in 2 additional index patients of a different genetic background, and emerges as a DUOX2 sequence segment prone to mutation. While the pathogenic impact of the 3 missense mutations and the spliced variant of DUOX2 was not formally ascertained in vitro, the fact that mutations correspond to highly conserved amino acids in vertebrate DUOX2 homologues and, mainly, because some perchlorate discharge tests were reportedly positive in family members harboring these mutations in simple heterozygosity make these novel defects likely to be causally linked to the hypothyroid phenotypes described below.

Vigone et al. [27] reported the first familial case of hypothyroidism associated with DUOX2 defect. Two Italian siblings harboring the same genotype (a *nonsense* R842X mutation and a *missense* R376W mutation in compound heterozygosity) showed relevant phenotypic differences. The first sibling was detected by TSH screening with a moderate but goitrous neonatal hypothyroidism (TSH: 173.2 mU/l). His brother was normal at the screening (TSH: 2.9 mU/l) but, because CH was detected in the first sibling, plasma TSH was determined, showing abnormal increases at 11 days (9.6 mU/l) and 45 days of life (18.4 mU/l). These important differences were attributed to the higher iodine load in the second sibling, as determined by urinary iodine. Both brothers were substituted with T_4 and after its withdrawal, both showed a mild but persistent elevation of TSH levels and PIOD (28 and 12% washout, respectively), for which thyroxine substitution was reintroduced. Perchlorate discharge in the euthyroid parents, respectively harboring each of the mutations, showed very mild but detectable PIOD (13 and 8% washouts). It cannot be excluded that the parents would have had a mild TCH described for monoallelic DUOX2 defects. This pedigree is interesting because, first, it suggests that DUOX2 phenotypes can be modified by environmental factors, such as individual iodine supplies. Second, it is also interesting because the environmentally mediated phenotypic variability of this DUOX2 defect prevented, in the case of the second sibling, the routine detection of the disease by neonatal screening, while biochemical hypothyroidism became clearly expressed after 1 month of age. The intellectual outcome of the brothers was assessed at T_4 withdrawal, with intellectual quotients (IQs) of 103 and 112, respectively.

Varela et al. [28] reported 2 families whose index patients had permanent hypothyroidism and PIOD. In family 1, the index patient was detected by screening. After re-evaluation at 5 years of age, hypothyroidism and goiter were confirmed and perchlorate discharge indicated PIOD (46% washout). This boy also had two DUOX2 mutations in compound heterozygosity: the described S965fsX994 and a novel missense mutation in the peroxidase domain (Q36H) (fig. 2a). While pathogenicity of the *frameshift* is clear, the impact of Q36H (Q36 being well-conserved among DUOX2 homologues) on DUOX2 function is less clear since 2 family members with the mutation in simple heterozygosity were reported to be euthyroid and to have a negative discharge test.

In family 2, 2 index patients presented severe neonatal CH and PIOD with 60–68% discharge. Late diagnosis and treatment of the first sibling led to mental retardation. Both are compound heterozygotes for two new DUOX2 variants, a clearly inactivating frameshift (G418fs482) with very early truncation of the protein before the first transmembrane domain (fig. 2a), and a nucleotide transversion in the acceptor site of intron 19 (g.IVS19-2A>C), which seems to affect exon splicing in minigene studies [28]. In vitro, g.IVS19-2A>C splices out exon 20, but authors also found skipping of exon 21 in both wild-type and mutant constructs, possibly representing an undescribed alternatively spliced mRNA of DUOX2. Worth noting is that if only exon 20 is skipped from DUOX2 cDNA because of the mutation, exon 21 would enter the sequence out of frame, leading to an early stop codon which would delete essential domains of the NADPH oxidase segment of DUOX2 and predict complete inactivation (fig. 2a). This severe pathogenicity neither seems in accordance with the PIOD determined in both index patients with 68 and 60% discharge, respectively, nor with the TIOD (as expected in individuals with biallelic and complete inactivation of DUOX2 [10]). On the other hand, g.IVS19-2A>C does have an impact on DUOX2 function, as indicated by the mild euthyroid hyperthyrotropinemia and PIOD (31% washout) identified in a child family member with this mutation in simple heterozygosity. Besides, when both exons 20 and 21 are spliced out, exon 22 enters the sequence in frame, encoding a protein with intact FAD-, NADPH-binding sites and transmembrane domains. The pathogenic impact of such a mutant would likely be less intense, and this would be more in agreement with the phenotypes mentioned. Further analyses are needed, e.g. retrotranscription of mRNA from blood lymphocytes, to confirm the precise splicing event taking place in these patients.

These two pedigrees from Argentina further reflect the interindividual phenotypic variability observed with the same DUOX2 genotypes, within a similar genetic background. Interestingly, the mild hypothyroidism present in an adult female of the second pedigree harboring a monoallelic, fully inactivating DUOX2 mutation (G418fs482) substantiates the notion of possible recurrence of hypothyroidism during adulthood in children who suffered from neonatal-infantile TCH.

Genotype-phenotype correlations of the 7 available familial pedigrees with DUOX2 defects, 4 described in the Netherlands and these 3 novel pedigrees from Italy and Argentina, suggest that permanency of hypothyroidism due to DUOX2 mutations seems to require alterations in both alleles. Patients with permanent hypothyroidism and DUOX2 mutations in compound heterozygosity have full inactivation of at least one DUOX2 allele by severe mutations (*nonsense* or frameshifts with premature truncation of the protein) accompanied by counterallele mutations that might allow some residual activity. TCH is so far associated with monoallelic severe inactivation of DUOX2. More patients' profiles might clarify in the future whether biallelic but very subtle defects of DUOX2 might also associate, under certain environmental conditions (e.g., iodine insufficiency) or life time events (e.g. immediate postnatal life or pregnancy), with transient hypothyroidism or whether, in contrast, monoallelic defects under e.g. (permanent) iodine insufficiency could behave as permanent phenotypes. The follow-up of 3 additional patients with DUOX2 mutations whose short evolution has not yet allowed classification of their hypothyroidism as permanent or transient will be helpful to analyze the effect of age on the expression of monoallelic DUOX2 defects in different geographical areas [29, 30].

While other patients with documented thyroid H_2O_2-generating defects are found in the literature [31, 32], no DUOX mutations have been studied and/or identified in these pedigrees. This, together with the absence of DUOX2, DUOX1 or TPO defects in 5 additional patients with TCH and PIOD included in the original study describing DUOX2 defects among Dutch patients, is compatible with the existence of other genetic factors involved in hydrogen peroxide production in the thyroid gland. This includes the novel maturation factor for DUOX2, DUOXA2, alterations of which have not been linked to any phenotype.

Only one case is reported in the literature that associates alteration in thyroidal H_2O_2 production and thyroid neoplasia [33]. H_2O_2 generation capacity of an excised 'cold nodule' tissue was found to be lost in comparison to surrounding tissue. This finding might not necessarily mean a direct involvement of H_2O_2 defects in the pathogenesis of the lesion of this patient (which was shown to be benign), but represents a secondary event framed in the known downregulation of markers of thyroid differentiation that occur in thyroid neoplasias. On the other hand, as previously reported, untreated dyshormonogenesis of the thyroid can result in thyroid tumors and malignancies under continuous, long-standing TSH stimulation [34–37]. This sequence of events is applicable to undiagnosed DUOX2 defects, only adding to the possible benefits of timely diagnosis and close follow-up of mild dyshormonogenetic defects of the thyroid.

No human disease associated with DUOX1 defects has been described. Based on the pattern of expression of the gene and on studies of DUOX genes in CH patients where DUOX1 was invariably found to be normal, it is tempting to speculate that putative DUOX1 phenotypes might be rather related to defective host defense against infection in the airways and other mucosal surfaces [17] than related to a thyroidal disorder.

Consequences of Transient Neonatal Hypothyroidism

As stated in the introduction, implementation of neonatal screening programs for CH has successfully eradicated the irreversible and severe neurological and psychomotor sequelae that used to devastate the lives of affected children with undiagnosed or late-diagnosed hypothyroidism in the prescreening era [2]. Despite this outstanding advance in the health care of the CH population and also the continued efforts to optimize screening programs and reduce to a minimum the time between birth, diagnosis and treatment of patients as well as to find optimal initial T_4 substitution dosages, studies on cognitive and motor outcomes of hypothyroid children and adolescents reveal the existence of persistent intellectual and motor deficits [38].

Because of its transitory nature, the consequences of TCH on psychomotor and intellectual outcome of children may have been underestimated in the past [39]. Even when TCH does not seem to affect parameters of physical development and psychomotor performance [40], accumulating evidence suggests that neonatal transient hypothyroidism and hyperthyrotropinemia are associated with impaired intellectual outcome [39, 41, 42]. This impairment is reflected in the loss of an average of 10 IQ points in children with TCH of 1–3 months' duration at 9 years of age compared to a matched control population [10]. The most affected IQ scores are the ones assessing the 'verbal' and the 'performance' skills as well as the global IQ. Most studies showing adverse long-term intellectual development of TCH children were performed in iodine deficiency areas, but their conclusions are applicable to any etiology of TCH.

Another problem is posed by the possibility that mildly hypothyroid children might 'escape' screening as false negatives, leaving us without the option of follow-up during infancy and childhood. A survey by Calaciura et al. [43] shows that 'subclinical' hypothyroidism in early childhood is a frequent outcome among children who were catalogued at screening as 'false negatives' because they had either completely normal (0.8–4.9 mU/l) or nearly normal/ borderline elevated (5–11.7 mU/l) TSH levels at confirmatory examination. The consequences of persistent hyperthyrotropinemia in children have not been studied, but in adults it is established that, even minimal abnormalities, may lead over the course of years to important or irreversible problems [43]. Lipid metabolism, myocardial function, linear growth and cognitive abilities are among the functions that may be adversely affected by 'subclinical hypothyroidism', which should be treated [44–46].

Concluding Remarks

DUOX1 and DUOX2, which are the components of the DUOX system of NADPH- and Ca^{2+}-dependent oxidases, are the latest members incorporated in the family of NOX proteins. While DUOX2 has been unequivocally associated with thyroid physiology through the production of H_2O_2 in the thyroid gland, indispensable for thyroid hormonogenesis, DUOX1 has not, and its precise role in thyroid and other tissues remains obscure.

Biallelic inactivation of DUOX2 in a patient with a complete block of thyroid hormone production reflects the importance of this protein for the synthesis of T_3 and T_4. Monoallelic inactivation of this gene has also been linked to milder and transient cases of neonatal hypothyroidism. While a possible recurrence of hypothyroidism coinciding with increases in T_4 requirement during some periods of life remains a theoretical possibility, the adult mild hypothyroidism shown in the newly described patients with the same genotype supports this idea. Additionally, compound heterozygosity for severe DUOX2 mutation and mutations with putative residual activity have been associated with permanent but mild ('subclinical') hypothyroidism. To what extent the clinical expression of this type of hypothyroidism through life depends on genetic determinants (number of alleles affected and severity of mutations), environmental factors (such as individual iodine supplies), physiological (such as puberty or pregnancy, which increase thyroid hormone demands) or pathological circumstances (such as thyroiditis, which reduces the capacity for thyroid hormone production) is a matter for further study, which has practical implications.

TCH is a frequent clinical situation. Optimization of screening programs to detect mild transient cases of hypothyroidism, sometimes taken as false positives at recall [43], is important, both for the negative impact that TCH has on the achievement of the full intellectual potential of children and to prevent 'subclinical hypothyroidism' in infancy and early childhood, and its consequences.

The discovery of the molecular basis of TCH just started with the identification of DUOX2 defects. A few reports suggest that a larger number of either known [47] or unknown genes might be involved in its pathogenesis. Clinical and molecular studies will now explore the possibility of genetic heterogeneity of TCH and be of help in the proper identification, molecular classification and management of patients with transient forms of hypothyroidism.

References

1 Vanderpump MPJ, Tunbridge WMG: The epidemiology of thyroid diseases; in Braverman LE, Utiger RD (eds): Werner & Ingbar's The Thyroid, a Fundamental and Clinical Text, ed 8. Philadelphia, Lippincott Williams & Wilkins, 2000, pp 467–473.

2 New England Congenital Hypothyroidism Collaborative: Effects of neonatal screening for hypothyroidism: prevention of mental retardation by treatment before clinical manifestations. Lancet 1981;ii:1095–1098.

3 Fisher DA, Dussault JH, Foley TP, Klein AH, LaFranchi S, Larsen PR, Mitchell ML, Murphey WH, Walfish PG: Screening for congenital hypothyroidism: results of screening one million North American infants. J Pediatr 1996;128:548–554.

4 Coackley JC, Francis I, Gold H, Mathur K, Connelly JF: Transient primary hypothyroidism in the newborn: experience of the Victorian Neonatal Thyroid Screening Program. Aust Paediatr J 1989; 25:25–30.

5 Vulsma T: Etiology and Pathogenesis of Congenital Hypothyroidism; thesis University of Amsterdam, 1991.

6 Léger J, Czernichow P: Hyperthyrotropinémie néonatale transitoire. Arch Fr Pediatr 1988;45: 783–786.

7 Vulsma T, de Vijlder JJM: Thyroid disease in newborns, infants and children; in Wass JA, Shalet SM (eds): Oxford Textbook of Endocrinology and Diabetes. New York, Oxford University Press, 2002, pp 532–544.

8 De Felice M, Di Lauro R: Thyroid development and its disorders: genetics and molecular mechanisms. Endocr Rev 2004;25:722–746.

9 Moreno JC, de Vijlder JJM, Vulsma T, Ris-Stalpers C: Genetic basis of hypothyroidism: recent advances, gaps and strategies for future research. Trends Endocrinol Metab 2003;14:318–326.

10 Moreno JC, Bikker H, Kempers MJ, van Trotsenburg AS, Baas F, de Vijlder JJ, Vulsma T, Ris-Stalpers C: Inactivating mutations in the gene for thyroid oxidase 2 (THOX2) and congenital hypothyroidism. N Engl J Med 2002;347:95–102.

11 Nose O, Harada T, Miyai K, et al: Transient neonatal hypothyroidism probably related to immaturity of thyroidal iodine organification. J Pediatr 1986;108:573–576.

12 Moreno JC, Keijser R, Gilhuijs-Pederson L, Gestel D, de Vijlder JJM, Ris-Stalpers C: Cloning and characterization of the human thyroid dehalogenase. Horm Res 2003;60(suppl 2):2.

13 Dupuy C, Ohayon R, Valent A, Noel-Hudson MS, Deme D, Virion A: Purification of a novel flavoprotein involved in the thyroid NADPH oxidase. Cloning of the porcine and human cDNAs. J Biol Chem 1999;274:37265–37269.

14 De Deken X, Wang D, Many MC, Costagliola S, Libert F, Vassart G, Dumont JE, Miot F: Cloning of two human thyroid cDNAs encoding new members of the NADPH oxidase family. J Biol Chem 2000;275:23227–23233.

15 Moreno JC, Pauws E, van Kampen AH, Jedlickova M, de Vijlder JJ, Ris-Stalpers C: Cloning of tissue-specific genes using serial analysis of gene expression and a novel computational substraction approach. Genomics 2001;75:70–76.

16 Edens WA, Sharling L, Cheng G, Shapira R, Kinkade JM, Lee T, Heather A, Edens HA, Tang X, Sullards C, Flaherty DB, Benian GM, Lambeth JD: Tyrosine cross-linking of extracellular matrix

is catalyzed by Duox, a multidomain oxidase/peroxidase with homology to the phagocyte oxidase subunit gp91phox. J Cell Biol 2001;154:879–891.
17 Ris-Stalpers C: Physiology and pathophysiology of the DUOXes. Antioxid Redox Signal 2006;8:1563–1572.
18 De Deken X, Wang D, Dumont JE, Miot F: Characterization of ThOX proteins as components of the thyroid H_2O_2 generating system. Exp Cell Res 2002;273:187–196.
19 Grasberger H, Refetoff S: Identification of the maturation factor for dual oxidase. Evolution of an eukaryotic operon equivalent. J Biol Chem 2006;281:18269–18272.
20 Grasberger H, De Deken X, Miot F, Pohlenz J, Refetoff S: Missense mutations of the dual oxidase 2 (DUOX2) implicated in congenital hypothyroidism have impaired trafficking in cells reconstituted with DUOX2 maturation factor. Mol Endocrinol DOI:10.1210/me.2007-0018.
21 Leusen JHW, Verhoeven AJ, Roos D: Interactions between the components of the human NADPH oxidase: intrigues in the *phox* family. Front Biosci 1996;1:d72–d90.
22 Wang D, De Deken X, Milenkovic M, Song Y, Pirson I, Dumont JE, Miot F: Identification of a novel partner of duox: EFP1, a thioredoxin-related protein. J Biol Chem 2005;280:3096–3103.
23 Donkó A, Péterfi Z, Sum A, Leto T, Geiszt M: Dual oxidases. Phil Trans R Soc B 2005;360: 2301–2308.
24 Harper RW, Xu C, McManus M, Heidersbach A, Eiserich JP: Duox2 exhibits potent heme peroxidase activity in human respiratory tract epithelium. FEBS Lett 2006;580:5150–5154.
25 Delange F, Bourdoux P, Ketelbant-Balasse P, Van Humskerken A, Glinoer D, Ermans AM: Transient primary hypothyroidism in the newborn; in Dussault JH, Walker P (eds): Congenital Hypothyroidism. New York, Marcel Dekker, 1983, pp 275–301.
26 Haddow JE, Palomaki GE, Allan WC, Williams JR, Knight GJ, Gagnon J, O'Heir CE, Mitchell ML, Hermos RJ, Waisbren SE, Faix JD, Klein RZ: Maternal thyroid deficiency during pregnancy and subsequent neuropsychological development of the child. N Engl J Med 1999;341:549–555.
27 Vigone MC, Fugazzola L, Zamproni I, Passoni A, di Candia S, Chiumello G, Persani L, Weber G: Persistent mild hypothyroidism associated with novel sequence variants of the DUOX2 gene in two siblings. Hum Mutat 2005;26:395.
28 Varela V, Rivolta CM, Esperante SA, Gruñeiro-Papendieck L, Chiesa A, Targovnik HM: Three mutations (p.Q36H, p.G418fsX482 and g.IVS19-2A>C) in the dual oxidase 2 gene responsible for congenital goiter and iodide organification defect. Clin Chem 2006;52:182–191.
29 Pfarr N, Korsch E, Kaspers S, Herbst A, Stach A, Zimmer C, Pohlenz J: Congenital hypothyroidism caused by new mutations in the dual oxidase 2 (THOX2) gene. Clin Endocrinol 2006;65: 810–815.
30 Di Candia S, Zamproni I, Cortinovis F, Passoni A, Vigone MC, Fugazzola L, Persani L, Weber G: Congenital hypothyroidism and partial iodide organification defects: two mutations in DUOX2 gene. Horm Res 2006;65(suppl 4):38.
31 Niepomniszcze H, Targovnik HM, Gluzman BE, Curuchet P: Abnormal H_2O_2 supply in the thyroid of a patient with goiter and iodine organification defect. J Clin Endocrinol Metab 1987;65: 344–348.
32 Figueiredo MLD, Cardose LC, Ferreira ACF, Campos DVB, da Cruz Domingos M, Corbo R, Nasciutti LE, Vaisman M, Carvalho DP: Goiter and hypothyroidism in two siblings due to impaired Ca^{2+}/NAD(P)H-dependent H_2O_2-generating activity. J Clin Endocrinol Metab 2001;94: 4843–4848.
33 Demester-Mikkine M, van Sande J, Corvilain J, Dumont JE: Benign thyroid nodule with normal iodide trap and defective organification. J Clin Endocrinol Metab 1975;41:1169–1171.
34 Vickery AL: The diagnosis of malignancy in dyshormonogenetic goitre. Clin Endocrinol Metab 1981;10:317–335.
35 Alabbasy AJ, Delbridge L, Cowell C, Silink M: Microfollicular thyroid adenoma and congenital goitrous hypothyroidism. Arch Dis Child 1992;67:1294–1295.
36 Camargo R, Limbert E, Gillam M, et al: Aggressive metastatic follicular thyroid carcinoma with anaplastic transformation arising from a long-standing goiter in a patient with Pendred's syndrome. Thyroid 2001;11:981–988.
37 Alzahrani AS, Baitei EY, Zou M, Shi Y: Clinical case seminar: metastatic follicular thyroid carcinoma arising from congenital goiter as a result of a novel splice donor site mutation in the thyroglobulin gene. J Clin Endocrinol Metab 2006;91:740–746.

38 Kempers MJE, van der Sluijs Veer L, Nijhuis-van der Sanden MWG, Kooistra L, Wiedijk BM, Faber I, Last BF, de Vijlder JJM, Grootenhuis MA, Vulsma T: Intellectual and motor development of young adults with congenital hypothyroidism diagnosed by neonatal screening. J Clin Endocrinol Metab 2006;91:418–424.
39 Rapaport R: Congenital hypothyroidism: expanding the spectrum. J Pediatr 2000;136:10–12.
40 Gruters A, Kohler B, Schnabel D, Helge H: Follow-up of thyroid function, thyroid size and development in children with transient neonatal hypothyroidism up to the age of 14 years. J Endocrinol Invest 1994;17(suppl 1):59.
41 Calaciura F, Mendorla G, Distefano M, Castorina F, Fazio T, Motta RM, Sava L, Delange F, Vigneri R: Childhood IQ measurements in infants with transient congenital hypothyroidism. Clin Endocrinol (Oxf) 1995;43:473–477.
42 Azizi F, Afkhami M, Sarshar A, Nafarabadi M: Effects of transient neonatal hyperthyrotropinemia on intellectual quotient and psychomotor performance. Int J Vitam Nutr Res 2001:71:70–73.
43 Calaciura F, Motta RM, Miscio G, Ficher G, Leonardi D, Carta A, Trischitta V, Tassi V, Sava L, Vigneri R: Subclinical hypothyroidism in early childhood: a frequent outcome of transient neonatal hyperthyrotropinemia. J Clin Endocrinol Metab 2002;87:3209–3214.
44 Surks MI, Ocampo E: Subclinical thyroid disease. Am J Med 1996;100:217–223.
45 Hack AE, Pols HA, Visser TJ, Drexhage HA, Hofman A, Witterman JC: Subclinical hypothyroidism is an independent risk factor for atherosclerosis and myocardial infarction in elderly women: The Rotterdam study. Ann Intern Med 2000;132:270–278.
46 McDermott MT, Ridgway EC: Subclinical hypothyroidism is mild thyroid failure and should be treated. J Clin Endocrinol Metab 2001;86:4591–4599.
47 Niu D-M, Lin C-Y, Hwang B, Jap T-S, Liao J-S, Wu J-Y: Contribution of genetic factors to neonatal transient hypothyroidism. Arch Dis Child Fetal Neonatal Ed 2005;90:F69–F72.

José C. Moreno, MD, PhD
Department of Internal Medicine, Erasmus Medical Center, Erasmus University
Dr. Molewaterplein 50
NL–3015 GE Rotterdam (The Netherlands)
Tel. +31 10 463 3385, Fax +31 10 463 5430, E-Mail j.morenonavarro@erasmusmc.nl

Van Vliet G, Polak M (eds): Thyroid Gland Development and Function.
Endocr Dev. Basel, Karger, 2007, vol 10, pp 118–126

Thyroid Hormone Transporter Defects

Annette Grüters

Institute for Experimental Pediatric Endocrinology, Charité Children's Hospital, Humboldt and Free University, Berlin, Germany

Abstract

In in vitro experiments, active transport of thyroid hormones had been repeatedly demonstrated. The membrane transporters for thyroid hormones which have been identified include the organic anion transporting polypeptide, heterodimeric amino acid transporters and the monocarboxylate transporters (MCT) which are the focus of this chapter. The gene encoding MCT8 which was identified as a specific thyroid hormone transporter is located on chromosome Xq13.2. The expression pattern of MCT8 indicates that MCT8 plays an important role in the development of the central nervous system by transporting thyroid hormone into neurons as its main target cells. Mutational analysis of the MCT8 gene revealed mutations or deletions in the MCT8 gene in unrelated male patients with severe psychomotor retardation and biochemical findings consistent with thyroid hormone resistance. Indeed, thyroid function tests in patients with MCT8 mutations demonstrated marked elevations of serum T_3 (in the thyrotoxic range), a significant decrease in serum T_4 or fT_4 and normal to elevated TSH levels.

Thyroid Hormone Transport into Cells

Thyroid hormones are important for the development of many tissues and their metabolic function throughout life. Of outmost importance is the critical role of thyroid hormones in brain and nervous system development [1, 2]. The biological effects of thyroid hormones are mediated through the binding of T_3 to nuclear receptors, which leads to a change in the interaction of the receptors with T_3-responsive elements in regulatory regions of many genes [3]. The thyroid produces mainly thyroxine (T_4), which is converted to T_3 through outer ring deiodination by the type 1 (D1) or type 2 (D2) iodothyronine deiodinase. In contrast, inner ring deiodination of T_4 to reverse T_3 (rT_3) or of T_3 to 3,3-diiodothyronine by the type 3 iodothyronine deiodinase (D3) leads to inactivation [4]. It has been assumed that hepatic D1 is the main source of serum T_3, whereas in the central nervous system (CNS) D2 is important for local regulation of thyroid hormone bioactivity. Since the T_3 receptors and the deiodinases are located in the cells, cellular entry is required for conversion of the thyroid hormones by intracellular deiodinases. Until recently it was assumed that the uptake of the lipophilic thyroid hormones into the cells was achieved by passive diffusion through lipid membranes. However, in in vitro experiments active transport of thyroid hormones had been repeatedly demonstrated [5, 6]. The membrane transporters for thyroid hormones which have been identified include the organic anion transporting polypeptide, heterodimeric amino acid transporters and the monocarboxylate transporters (MCT).

Monocarboxylate Transporters

To date, 14 members of the MCT family have been identified in various tissues from different species. The MCTs are proteins of 426–565 amino acids with 12 predicted transmembrane domains with the N- and the C-terminal domains of the proteins located inside the cells. Members of the MCT family transport monocarboxylates such as lactate, pyruvate, and ketone bodies. For expression at the plasma membrane, MCTs require ancillary proteins. CD147 is a widely distributed cell surface glycoprotein, which enables the proper expression and function of MCTs at the cell surface [7]. To date 14 different genes encoding MCTs are known, which are located on several human chromosomes (see table 1). All MCTs have a widely distributed expression, some are ubiquitously expressed and several MCTs show a neuronal expression.

The gene encoding MCT8 is located on chromosome Xq13.2. It consists of 6 exons and codes for a protein of about 67 kDa [8]. The protein contains 12 predicted transmembrane domains and contains at the N-terminus, a PEST domain rich in Pro (P), Glu (E), Ser (S), and Thr (T) residues (fig. 1). PEST domains serve as proteolytic signals, which result in rapid degradation of the protein and a reduced half-life. Expression of MCT8 was described in the human heart, brain, placenta, lung, kidney, skeletal muscle and in the liver. No biological function of MCT8 was known until Friesema et al. [9] identified MCT8 as a specific thyroid hormone transporter.

It was shown that expression of MCT8 induced an approximate 10-fold increase in uptake of T_4 and T_3 relative to control oocytes. MCT8 transports T_4, T_3, rT_3, and T_2, but not sulfonated iodothyronines, the amino acids Tyr, Trp, Leu, and Phe, and lactate. There is a high degree of homology in the amino sequences of rat, mouse, and human MCT8.

MCT8 is expressed in numerous human tissues, including brain, heart, placenta, lung, kidney, skeletal muscle and liver. Detailed studies of MCT8

Table 1. Clinical and biochemical findings in patients with MCT8 mutations, patients with Allan-Herndon-Dudley syndrome and in mice with targeted deletion of MCT8

Clinical symptoms	Patients with MCT8 mutations	Patients with Allan-Herndon-Dudley syndrome	MCT8 knockout mice
Generalized muscular hypotonia in infancy	+	+	absent
Increasing spasticity with age	+	+	absent
Spastic quadriplegia	+	+	absent
Involuntary dystonic movements	+	+	absent
Nystagmus	+	+	not reported
Vision	normal	normal	normal
Secondary microcephalus	+	+	–
Feeding problems	+	+	absent
Sleeping disorder	+	not reported	absent
Seizures	+	(+)	absent
Mental retardation	+	+	–
Ability to walk	absent	+	+
Ataxia	–	+	–
Speech development	absent	+	not applicable
Dysarthric speech	–	+	not applicable
Facial abnormalities (elongated face, depressed nasal bridge)	–	+	absent
Increased serum T_3	+	+	+
Low serum T_4	+	+	+
Normal/increased TSH	+	+	+
Increased D1/D2 activity	+	not reported	+

expression in the CNS have been performed in mice [10]. Highest transcript levels for MCT8 were detected in the choroid plexus of the lateral, third, and fourth ventricles. Significant MCT8 mRNA concentrations were also found in the amygdala, in the pyramidal cell layer of the hippocampal formation and in the granule cell layer of the dentate gyrus. Hybridization was also found throughout striatal areas, with high transcript levels in the olfactory tubercle and to a lesser extent in the caudate-putamen and nucleus accumbens. Lower expression signals were detected in the striatum and cerebellum. MCT8 is mainly expressed in neurons, in which coexpression of D2 could be demonstrated [10]. This expression pattern of MCT8 indicates that MCT8 plays an important role in the development of the CNS by transporting thyroid hormone into neurons as its main target cells.

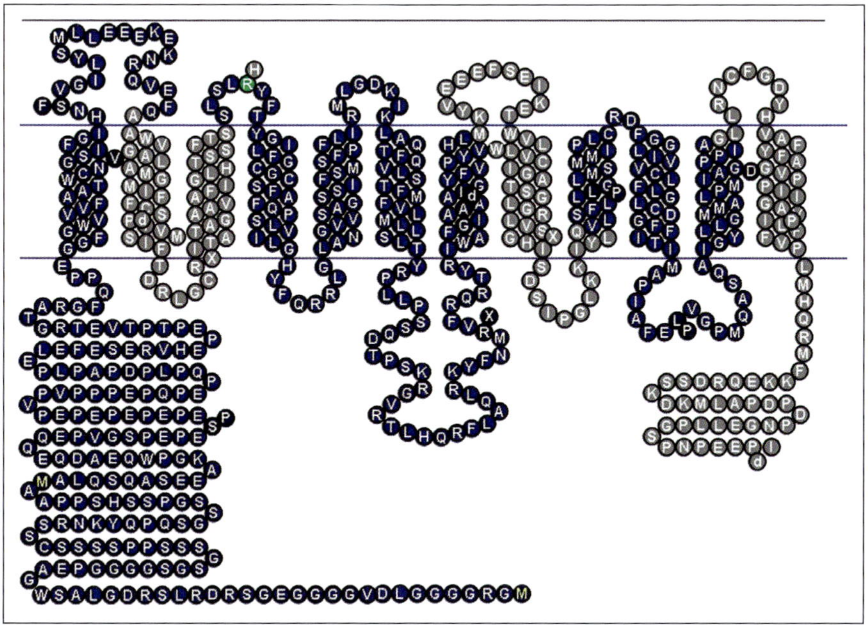

Fig. 1. Putative structure of the MCT8 protein.

MCT8 mRNA was detected also in early first trimester placenta with a significant increase with advancing gestation. MCT8 immunostaining was demonstrated in villous cytotrophoblast, syncytiotrophoblast and extravillous trophoblast cells with increasing intensity with advancing gestation. It has been concluded that the expression of MCT8 in placenta from early gestation is compatible with an important role in thyroid hormone transport during fetal development and a specific role in placental development [11].

Human Phenotype of MCT8 Mutations

Since the MCT8 gene is located on the X chromosome, it was suggested that mutations in MCT8 could cause an X-linked form of thyroid hormone resistance and mental retardation. Subsequently, mutational analysis of the MCT8 gene revealed mutations or deletions in the MCT8 gene in unrelated male patients with severe psychomotor retardation and biochemical findings consistent with thyroid hormone resistance [12–14]. Until the mutation was identified in all these boys the cause of the retardation had been unknown. In the mothers of all patients the mutations were detected as well, but as expected for an X-linked disorder, none of the mothers had signs of mental retardation or impairment of motor development.

More recently, MCT8 gene mutations have been found to be the cause of Allan-Herndon-Dudley syndrome, an X-linked syndromic form of mental retardation first described in 1944 [15] (Online Mendelian Inheritance in Man, access No. OMIM 309600).

Clinical Findings

The clinical findings in most patients with MCT8 mutations are uniform: in the first year of life general hypotonia with a complete inability to hold the head, lack of eye fixation and involuntary movements is observed. Later on, the inability to sit and walk as well as involuntary dystonic movements of the arms and spontaneous activity of facial muscles are observed. Primitive (glabella, snout) reflexes are easily elicited or occur spontaneously as well as bilateral asymmetric tonic neck reflex.

Nystagmus is frequently present, but hearing and vision are found to be normal, as demonstrated by brainstem evoked response audiometry. Microcephaly may occur, but it is possible that this is a secondary event. Generalized epilepsy with tonic and tonic-clonic seizures leads to antiepileptic drug treatment in many patients. In some patients, mental development may improve with age, since smiling, laughing as well as voluntary grasping is present in older patients. Most likely as a consequence of the neurological impairment, severe feeding problems and failure to thrive are a frequent complication sometimes requiring a gastrostomy feeding tube. Sleeping disorders are also a frequent finding and speech development is absent in all patients. Spastic quadriplegia has developed in some of the patients.

Allan-Herndon-Dudley syndrome was one of the first described syndromes of X-linked mental retardation. In six large families linkage studies have mapped the gene locus to Xq13.2 and after the identification of the human MCT8 mutations described above, mutations have been found in each of the six families. There are some differences with the clinical findings described above. Some of these patients learned to walk independently, but exhibited severe ataxia, which ultimately made them wheelchair-bound. Patients who learned to speak exhibited dysarthric speech. There is a transition in childhood to spasticity manifested by hyperreflexia at the large joints, clonus, and a positive Babinski sign.

In addition to the neurological findings comparable to those described above, facial manifestations with bifrontal narrowing, simply formed or cupped ears, elongated and myopathic faces with midface hypoplasia, narrow high palate, anteverted nares, spaced teeth and square faces were described.

The development of all 32 affected males who were evaluated was severely impaired from birth. Compatibility with longevity was evident from the 8 affected males, of the 32 that were evaluated, who lived beyond 70 years. Since most affected males were unable to meet the baseline of standardized cognitive tests, the precise level of mental retardation could not be determined.

Prenatal and postnatal growth as well as pubertal development seem to be grossly normal. Other clinical symptoms observed in congenital hypothyroidism, e.g. prolonged jaundice, constipation, bradycardia, delayed skeletal maturation or skin manifestations such as myxedema, have never been reported.

Thyroid Function Tests

Thyroid function tests in the patients with MCT8 mutations demonstrated marked elevations of serum T_3 (in the thyrotoxic range), a significant decrease in serum T_4 or fT_4 and normal to elevated TSH levels, compatible with a diagnosis of apparent thyroid hormone resistance. rT_3 levels are normal or decreased.

When measured, serum sex hormone-binding globulin was found to be significantly elevated, indicating a hyperthyroid state of the liver.

In some of the patients treatment with high doses of L-thyroxine suppressed serum TSH, but had no clinical effects. In contrast, in 1 patient treatment with T_3 alone (20 μg/day) induced a decrease in serum TSH and fT_4 and treatment with T_4 (100 μg/day) plus T_3 (30 μg/day) resulted in impaired weight gain and increased sweating, but did not improve neurological symptoms [16].

From these manifestations, it can be concluded that mutations in MCT8 predominantly result in a severe neurological phenotype, because MCT8 is essential for the proper local supply of T_3 to neurons of the CNS avoiding lack or excess of thyroid hormones. However, it still cannot be excluded that MCT8 might also be involved in the transport of other unidentified substances.

Since typical clinical symptoms of hypothyroidism are lacking in other organ systems impaired MCT8 function seems to cause tissue-specific hypothyroidism in the (developing) CNS. The lack of manifestations in other organs therefore implies that there are other compensating thyroid hormone transport mechanisms in these tissues. However, since besides the liver, no other organ, e.g. the intestinal tract and the heart, exhibits hyperthyroid symptoms, whether enhanced local degradation of thyroid hormones is present in these tissues remains to be established.

Molecular Findings

In figure 2, the mutations which have been identified so far are depicted in red. In vitro functional studies with transfected mutant constructs of the missense mutations revealed the absence of T_3 uptake for most of the mutants. Only the transfection of the Arg271His mutation resulted in a residual T_3 uptake, which was however significantly reduced compared with transfection of the

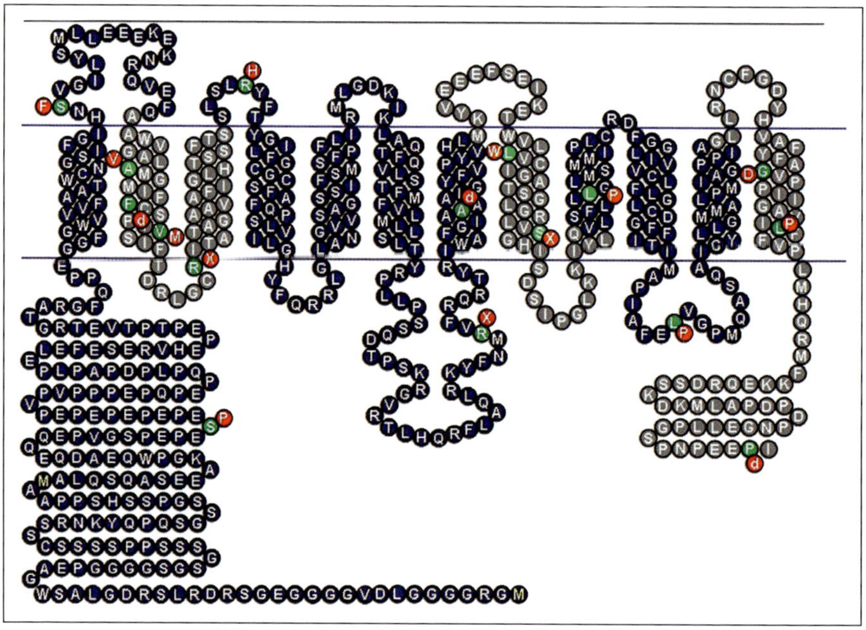

Fig. 2. Human mutations of the MCT8 protein.

wild-type (WT) MCT8. Furthermore, with the Ala150Val mutant, a reduced cell surface expression of the mutated transporter in comparison to WT MCT8 was demonstrated. The subcellular localization of the mutated transporter revealed a more reticular expression pattern than expression on the cell membrane, which could be clearly demonstrated for the WT MCT8. Moreover, it was shown that the Ala150Val mutant MCT8 forms homodimers and that dimerization is comparable to the WT MCT8 in vitro [16].

Studies of cultured skin fibroblasts isolated from the patients demonstrated a decreased uptake of both T_4 and T_3 [17] as well as a significantly increased D2 activity. The increased conversion of T_4 to T_3 by D2 together with a decreased T_3 uptake may lead to the decreased serum T_4 and increased serum T_3 concentrations.

Animal Model: Targeted Disruption of MCT8 in Mice

As described by Dumitrescu [17] and according to known observations in a different knockout approach, deletion of the *MCT8* gene in mice leads to biochemical findings comparable to those observed in human patients. The MCT8 knockout mice generated by Dumitrescu et al., like the human patients, present with low T_4 and high T_3 serum concentrations and pituitary resistance to thyroid hormone. Also in the animal model increased D1 and D2 activities were observed in several tissues including the CNS.

In spite of the fact that *MCT8* deletion reproduced the biochemical phenotype in mice, no obvious signs of neurological disturbances were observed: the mice did not present with abnormalities in motor development or behavior. The crucial questions to be addressed in further research studies are the potential compensating mechanisms of thyroid hormone transport in rodents, e.g. by other transporters or different sensitivities of developing structures of the CNS to hyper- or hypothyroidism. The lack of MCT8-mediated thyroid hormone transport in MCT8 knockout mice is not comparable to the situation in $Pax8^{-/-}$ mice, because residual thyroid hormone function may be present due to alternative transport mechanisms.

Concluding Remarks

Hemizygous MCT8-deficient males present a syndrome with two components: a thyroid defect (increased total and free serum T_3 and decreased total and free T_4 and rT_3 concentrations) and severe psychomotor and developmental delay (generalized dystonia combined with spasticity, mental retardation, lack of verbal communication, poor head control and coordination). As is the case for most X-linked diseases, males are more severely affected in terms of both the neurological and thyroid defects, whereas female carriers have only mild thyroid function test abnormalities. Since MCT8 is not the only thyroid hormone transporter, different human tissues may express different transporters. Mutations in MCT8 will cause local hypothyroidism in tissues depending on MCT8, possibly in different regions of the developing CNS. Concomitantly, the high serum T_3 concentrations may cause a hyperthyroid state of tissues independent of MCT8, e.g. the liver. Therefore, the differences between the human and rodent phenotype may reflect differences in the effect of hypo- and hyperthyroid states of various tissues, especially of the developing CNS.

References

1 Oppenheimer JH, Schwartz HL: Molecular basis of thyroid hormone-dependent brain development. Endocr Rev 1997;18:462–475.

2 Morreale de Escobar G, Obregón MJ, Escobar del Rey F: Role of thyroid hormone during early brain development. Eur J Endocrinol 2004;151:U25–U37.

3 Bernal J: Action of thyroid hormone in brain. J Endocrinol Invest 2002;25:268–288.

4 Bianco AC, Salvatore D, Gereben B, Berry MJ, Larsen PR: Biochemistry, cellular and molecular biology, and physiological roles of the iodothyronine selenodeiodinases. Endocr Rev 2002;23: 38–89.

5 Hennemann G, Docter R, Friesema EC, de Jong M, Krenning EP, Visser TJ: Plasma membrane transport of thyroid hormones and its role in thyroid hormone metabolism and bioavailability. Endocr Rev 2001;22:451–476.

6 Friesema EC, Jansen J, Milici C, Visser TJ: Thyroid hormone transporters. Vitam Horm 2005;70: 137–167.

7 Kirk P, Wilson MC, Heddle C, Brown MH, Barclay AN, Halestrap AP: CD147 is tightly associated with lactate transporters MCT1 and MCT4 and facilitates their cell surface expression. EMBO J 2000;19.3896–3904.

8 Lafrenière RG, Carrel L, Willard HF: A novel transmembrane transporter encoded by the XPCT gene in Xq13.2. Hum Mol Genet 1994;3:1133–1139.

9 Friesema EC, Ganguly S, Abdalla A, Manning Fox JE, Halestrap AP, Visser TJ: Identification of monocarboxylate transporter 8 as a specific thyroid hormone transporter. J Biol Chem 2003;278: 40128–40135.

10 Heuer H, Maier MK, Iden S, Mittag J, Friesema EC, Visser TJ, Bauer K: The monocarboxylate transporter 8 linked to human psychomotor retardation is highly expressed in thyroid hormone-sensitive neuron populations. Endocrinology 2005;146:1701–1706.

11 Chan S-Y, Franklyn JA, Pemberton HN, Bulmer JN, Visser TJ, McCabe CJ, Kilby MD: Monocarboxylate transporter 8 expression in the human placenta: the effects of severe intrauterine growth restriction. J Endocrinol 2006;189:465–471.

12 Friesema EC, Grueters A, Biebermann H, Krude H, von Moers A, Reeser M, Barrett TG, Mancilla EE, Svensson J, Kester MH, et al: Association between mutations in a thyroid hormone transporter and severe X-linked psychomotor retardation. Lancet 2004;364:1435–1437.

13 Dumitrescu AM, Liao XH, Best TB, Brockmann K, Refetoff S: A novel syndrome combining thyroid and neurological abnormalities is associated with mutations in a monocarboxylate transporter gene. Am J Hum Genet 2004;74:168–175.

14 Lenzner S, Rosenkranz MD, Grueters A, Biebermann H, Chelly J, Moraine C, Frijns J, van Bokhoven H, Visser T, Ropers H: Severe X-linked mental retardation caused by mutations in the gene for the thyroid hormone transporter MCT8. European Society of Human Genetics Concurrent Symposia, Munich, 2004.

15 Schwartz CE, May MM, Carpenter NJ, Rogers RC, Martin J, Bialer MG, Ward J, Sanabria J, Marsa S, Lewis JA, et al: Allan-Herndon-Dudley syndrome and the monocarboxylate transporter 8 (MCT8) gene. Am J Hum Genet 2005;77:41–53.

16 Biebermann H, Ambrugger P, Tarnow P, von Moers A, Schweizer U, Grueters A: Extended clinical phenotype, endocrine investigations and functional studies of a loss-of-function mutation A150V in the thyroid hormone specific transporter MCT8. Eur J Endocrinol 2005;153:359–366.

17 Dumitrescu AM, Liao XH, Weiss RE, Millen K, Refetoff S: Tissue-specific thyroid hormone deprivation and excess in monocarboxylate transporter (Mct) 8-deficient mice. Endocrinology 2006;147:4036–4043.

Prof. Annette Grüters
Institute for Experimental Pediatric Endocrinology
Charité Children's Hospital, Humboldt and Free University
Augustenburger Platz 1, DE–13353 Berlin (Germany)
Tel. +49 30 450 566 251, Fax +49 30 450 566 936, E-Mail annette.grueters@charite.de

Van Vliet G, Polak M (eds): Thyroid Gland Development and Function.
Endocr Dev. Basel, Karger, 2007, vol 10, pp 127–139

Novel Biological and Clinical Aspects of Thyroid Hormone Metabolism

Alexandra M. Dumitrescu[a], *Samuel Refetoff*[a–c]

Departments of [a]Medicine and [b]Pediatrics and [c]Committee on Genetics, University of Chicago, Chicago, Ill., USA

Abstract

Intracellular metabolism of thyroid hormone (TH) and availability of the active hormone T_3 is regulated by three selenoprotein iodothyronine deiodinases (Ds). D1 and D2 convert the precursor T_4 into the active hormone, T_3. D3 is the principal inactivator of T_4 and T_3 to their respective metabolites, rT_3 and T_2. While acquired changes in D activities are common, inherited defects in humans were not known. Recently, two families with abnormal thyroid function tests, high serum T_4, high rT_3, low T_3 and slightly increased TSH, were identified. Linkage analysis and sequencing excluded abnormalities in all 3 DIO genes, yet clinical studies showed reduced responsiveness to T_4 but not to T_3. Extensive search for putative defects in genes involved in D2 metabolism led to the identification of mutations in the Sec insertion sequence binding protein (SBP)2 gene, involved in the synthesis of selenoproteins, including Ds. Affected children were either homozygous or compound heterozygous for these mutations. Other selenoproteins, including glutathione peroxidase, were also reduced in affected subjects, confirming a generalized effect of the SBP2 defect. Opposite thyroid test abnormalities are found in mutations of the TH transporter MCT8, and appear to be caused by the resulting increases in D2 and D1 activities.

Introduction and Physiologic Background

Thyroid hormones (TH) are iodinated compounds known to influence gene expression and metabolism in tissues of all vertebrate animals. The effects of TH are dependent on the quantity of the hormone that reaches peripheral tissues, their intracellular availability and the presence of unaltered TH receptors. Entry of TH into cells is an active process and several classes of TH membrane transporters with different kinetics and substrate preferences have been identified [1].

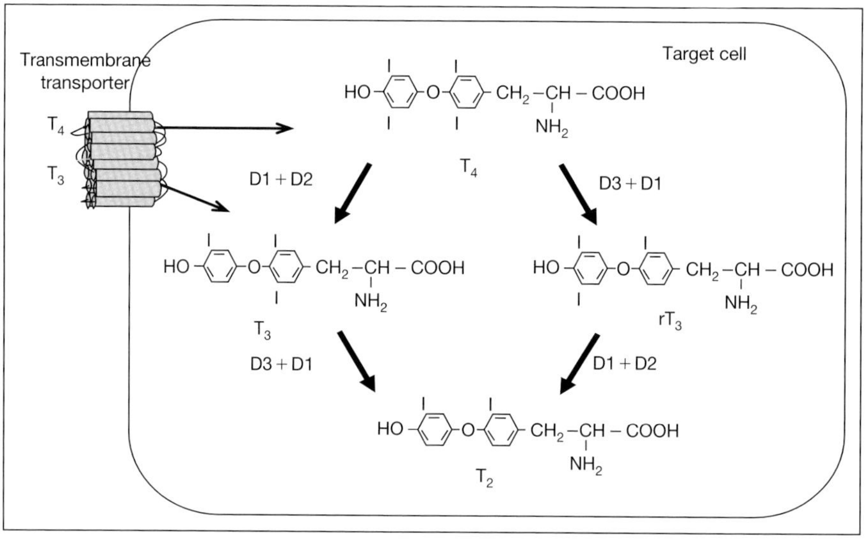

Fig. 1. Regulation of intracellular TH bioactivity. After active cellular uptake of TH through transmembrane transporters, the precursor 3,3′,5,5′-tetraiodothyronine (thyroxine, T_4) is converted into the active 3,3′,5-triiodothyronine (T_3) hormone or inactive 3,3′,5′-triiodothyronine metabolite (rT_3). D1 and D2 are the principal enzymes that catalyze 5′-deiodination, converting T_4 to T_3 and rT_3 to T_2, while D3 and to a lesser extent D1 catalyze 5-deiodination, converting T_4 to rT_3 and T_3 to T_2.

Once intracellularly present, the hormone precursor 3,3′,5,5′-tetraiodothyronine (thyroxine, T_4) is metabolized by removal of the outer ring iodine (5′-deiodination) to form 3,3′,5-triiodothyronine (T_3) or inactivate T_4 and T_3 by inner ring, 5-deiodination to form 3,3′,5′-triiodothyronine (reverse T_3, rT_3) and 3,3′-diiodothyronine (T_2), respectively [2]. The activating deiodinases are D1 and D2, while the inactivating enzyme is D3 and to a lesser extent D1 (fig. 1). Their presence in changing concentrations in various cell types allows an additional regulation of hormone supply at the cell level [2].

While genetic defects in thyroid gland development, TH synthesis, secretion and action have been identified, until recently, inherited defects in TH metabolism were not known [3]. Deiodinases are selenoenzymes requiring the presence of the rare amino acid selenocysteine (Sec) in the active center. Several factors are required for Sec incorporation: *cis*-acting sequences present in the mRNA of a selenoprotein [UGA codon and Sec insertion sequence (SECIS)] and *trans*-acting factors [elongation factor (eEF^{Sec}), $tRNA^{Sec}$, and SECIS-binding protein (SECISBP2 or SBP2)] [4] (fig. 2). Sec is the 21st amino acid and is structurally identical to cysteine (Cys), except that it contains

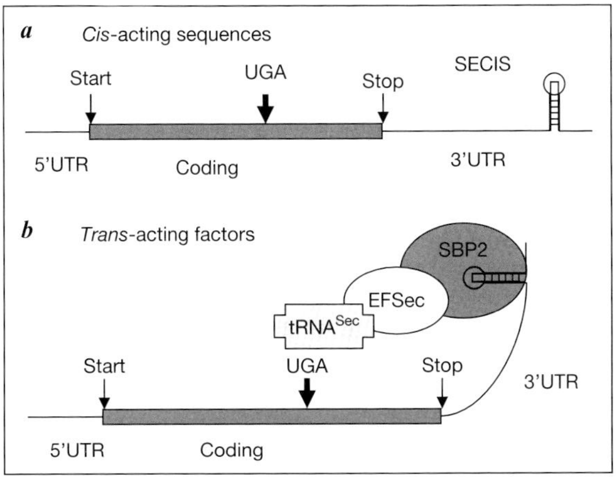

Fig. 2. Components involved in Sec incorporation. ***a*** *Cis*-acting sequences present in the mRNA of selenoproteins: an in-frame UGA codon and Sec incorporation sequence (SECIS) element, a stem loop structure located in the 3′UTR (untranslated region). ***b*** Binding of the *trans*-acting factor SECIS-binding protein (SBP2) which recruits the Sec-specific elongation factor (EFSec) and Sec-specific tRNA ($tRNA^{Sec}$). The result is a recoding of the UGA codon and Sec incorporation.

selenium instead of sulfur. Sec has a distinct functional advantage at physiological pH. When Sec is replaced with Cys, the catalytic activity of a selenoenzyme is drastically reduced [5].

Sec is incorporated through recoding of a UGA codon present in the mRNAs of a selenoprotein, dictated by the presence of SECIS, a characteristic stem-loop RNA structure in the 3′ untranslated region. Using the SECIS element as bait, the rat SECIS-binding protein, SBP2, was purified and cloned in 2000 [6]. We have recently identified a genetic defect in SBP2 that results in deficient deiodinase enzymatic activity perturbing the TH metabolism that manifested as a novel thyroid phenotype.

Clinical Presentation

Several members of two families presented with similar thyroid function test (TFT) abnormalities, namely, high T_4, low T_3, high rT_3 and slightly elevated TSH.

One family of Bedouin origin from Saudi Arabia (family A) was referred because of abnormal TFTs in three of seven siblings (fig. 3a). The propositus,

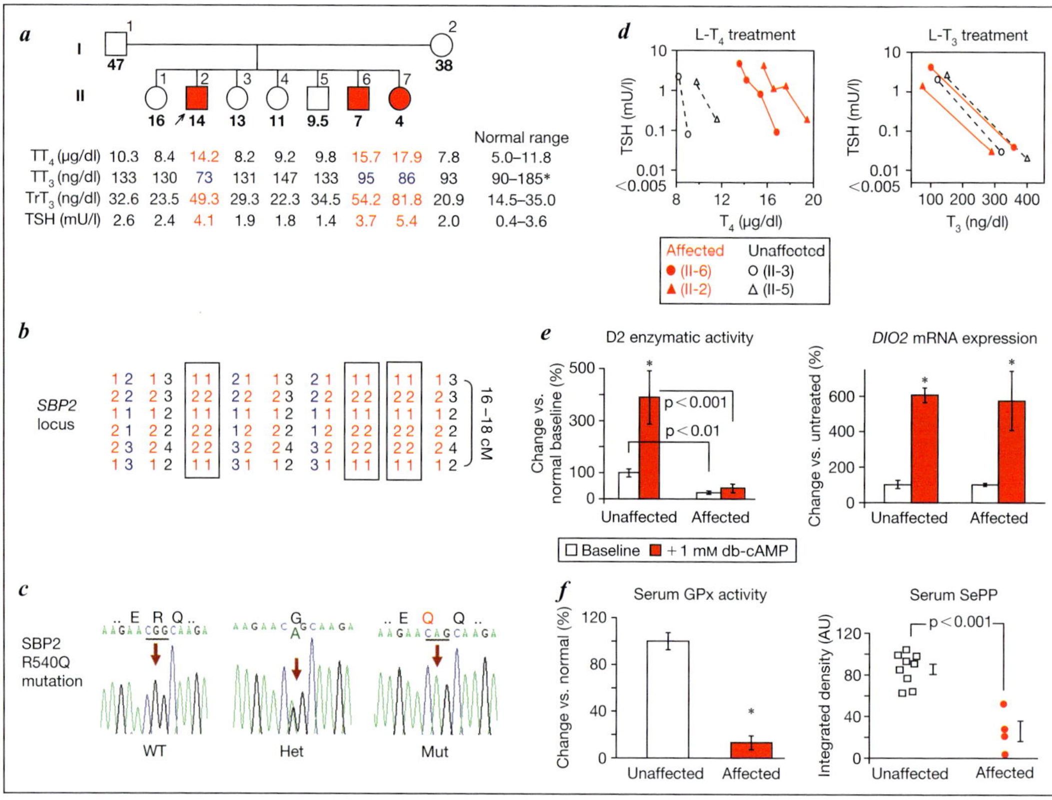

Fig. 3. Identification of a SBP2 gene mutation and the resulting in vivo and in vitro abnormalities. ***a*** Pedigree and TFTs in members of family A. TFT values are aligned under each subject symbol. TT_4 = Total T_4; TT_3 = total T_3; TrT_3 = total reverse T_3. Abnormal values are in bold characters, high values in red, low values in blue. * = The normal range for children is 100–205 μg/dl. ***b*** Haplotyping with genetic markers overlapping the SBP2 locus. Affected share homozygous haplotypes while the unaffected parents and siblings are heterozygous. ***c*** Sequences of the SBP2 region harboring the R540Q mutation. Arrow indicates the location of the G → A transition in a CpG dinucleotide changing the codon CGG to CAG. Sequences are those of an unaffected control (WT), a heterozygous parent (Het) and an affected homozygous subject (Mut). Corresponding amino acids are written above the nucleotide triplet. ***d*** Serum TSH and corresponding serum T_4 and T_3 levels, before and during the oral administration of incremental doses of L-T_4 and L-T_3. Note the higher concentrations of T_4 required to reduce serum TSH in the affected subjects. ***e*** Deiodinase 2 enzymatic activity and mRNA expression. Bars indicate ± SEM. Baseline and stimulated D2 activity is significantly lower in affected. There is a significant increase of DIO2 mRNA with db-cAMP, in both normals and affected (*$p < 0.001$). There are no significant differences in baseline and db-cAMP-stimulated DIO2 mRNA in affected versus the unaffected. ***f*** Effect of SBP2 deficiency on other selenoproteins. Bars indicate ± SEM. GPx enzymatic activity in

the second child born to unrelated parents from the same tribe, was 14 years old when he presented with short stature. Since age 11 he had been growing below the 3rd percentile and bone ages were 7 and 9 at chronological ages of 11 and 13, respectively, while still prepubertal [7]. Somatomedin C and stimulated growth hormone values were within the normal range. It is of note that, at neonatal screening, his serum TSH was high but no treatment was given as serum T_4 was not low. TSH values remained above normal but development proceeded normally. There was no hearing impairment as assessed by an audiogram. The concentrations of total T_4, free T_4 and total rT_3 were high while that of total and free T_3 were low. The same pattern of TFT abnormalities was found in a 7-year-old brother and 4-year-old sister (fig. 3a), both clinically euthyroid. TFT of all other family members, including parents and 4 siblings, were normal.

In the second family (family B) of mixed Irish and African origin, the single affected child presented at age 6 years with growth retardation and thyroid abnormalities similar to the children of family A. The mother mentioned frequent fainting after exertion but clinical, biochemical and electrophysiological investigations did not detect abnormalities.

In vivo and in vitro Studies

Defects in serum and transmembrane [8, 9] transport of TH were excluded. The potency of L-T_4 compared to L-T_3 to suppress TSH was used to test the hypothesis of a defect in deiodinase-mediated iodothyronine metabolism. When incremental doses of L-T_4 and L-T_3 were given to 2 affected and 2 normal siblings, higher doses and serum concentrations of T_4, but not T_3, were required to reduce TSH in the affected siblings, suggesting a defect in T_4 to T_3 conversion (fig. 3d). However linkage of the phenotype to loci of the three deiodinases was excluded and their sequences were normal.

Cultured skin fibroblasts were used to test in vitro the hypothesis of abnormal T_4 to T_3 conversion. Because D2, an integral membrane ER-resident selenoenzyme [2], but no D1 or D3 is expressed in these cells, all subsequent studies measured D2 expression and function. As the DIO2 gene contains a cAMP-responsive element (CRE), the synthetic analogue dibutyryl-cAMP (db-cAMP) also stimulates D2 expression and activity. The baseline D2 activity in

serum from all 4 affected children and 10 unaffected individuals. Results are expressed as percent of the average of unaffected subjects (*$p < 0.001$). SePP levels in serum from all 4 affected and available unaffected members of both families were quantified by scanning the Western blot. AU = Arbitrary units.

fibroblasts from the affected subjects was reduced to near or below the limit of detection (fig. 3e). In all fibroblasts from normal individuals, DIO2 mRNA and enzymatic activity increased with db-cAMP stimulation, whereas in the affected subjects, the increase in DIO2 mRNA was not accompanied by a corresponding increase in enzymatic activity (fig. 3e). These results further support the clinical findings of defective TH metabolism and provided direct evidence for abnormal D2 function.

SBP2 Gene Mutations

A posttranscriptional defect through impaired D2 metabolism or synthesis was considered as possibly responsible for the reduced enzymatic activity. Supply of active D2 enzyme is regulated by degradation in proteosomes through ubiquitination [10] and by deubiquitination [11]. Linkage was excluded for genes encoding proteins involved in D2 ubiquitination (*UbE2G1, UbE2G2, UbE2L3*) [10] and in D2 de-ubiquitination (*VDU1, VDU2*) [11]. From the components of the Sec incorporation machinery [4], a defect in *cis* sequences, $tRNA^{Sec}$ and EFSec was excluded by linkage analysis or by sequencing. However, affected subjects shared homozygous haplotypes at the SBP2 locus (fig. 3b). Sequencing revealed a missense mutation in exon 12, R540Q. Affected children were homozygous and the parents were heterozygous carriers (fig. 3c). The haplotype harboring the mutation is most likely inherited identical by descent in this Bedouin family. Extensive genotyping with markers flanking the locus showed a 16- to 18-cM segment shared identical by descent by the parents, suggestive of a recent common ancestor.

Mutations in the SBP2 gene were also found in family B, the affected child being compound heterozygous. He inherited from his father a nonsense mutation (K438X) and from the mother, an intronic mutation (IVS8ds+29 G $\rightarrow$ A). The latter mutation creates an alternative donor splice site producing an alternative transcript that incorporates 26 bp into exon 8. The abnormally spliced transcripts represented 52% of the transcripts generated from the mutant maternal allele in lymphocytes and change the reading frame to produce a putative truncated protein.

Population screens showed that these SBP2 gene sequence differences were not polymorphic in the respective populations. The homozygous mutation R540Q in family A is a nonsynonymous change located in a conserved amino acid across species and likely creates a hypomorphic rather than a null allele. In family B, IVS8ds+29 G $\rightarrow$ A results in partial alternative splicing and abnormal transcripts. The total amount of normal transcripts in the affected child was estimated to be 24%. In both instances the result is a partial, rather than a complete defect predicting a reduced rather than complete deficiency of selenoprotein synthesis and a mild phenotype.

Consequences for Other Selenoproteins

As SBP2 is epistatic to selenoprotein synthesis, identification of decreased D2 activity due to recessive SBP2 defect was likely to affect other selenoproteins. As D1 and D3 are too low to measure or absent in skin fibroblasts and lymphocytes, other selenoproteins were assessed. Glutathione peroxidase (GPx) activity was 7.5-fold lower in serum (fig. 3f) and 3.3-fold lower in fibroblasts of the affected as compared to normal subjects. Furthermore, selenoprotein P (SePP) levels in serum were significantly lower in the affected of both families compared with the unaffected (fig. 3f).

SBP2 is expressed at low levels in all tissues tested, with a high expression and an additional transcript in testis [12]. Human SBP2, cloned in 2002 [12], has 854 amino acids. The C-terminal domain is sufficient for all known functions of SBP2 including SECIS binding, ribosome binding and Sec incorporation [13]. A unique N-terminal domain contains a strong nuclear localization signal but its precise function is unknown [12].

How could a gene involved in the synthesis of an entire class of proteins manifest only as abnormal TFT? There appears to be a hierarchical preservation of selenoproteins during Se deprivation, conserving the enzymatic activity of those with a more important function [13, 14]. This hierarchy is supposedly produced by the rates of selenoprotein degradation and by the functional demands of particular selenoproteins. Therefore, selenoproteins with short half-life and high demand for their function, such as D2, might be the first to fail when the Sec incorporation machinery becomes inefficient.

Other Inherited Defects That Affect Iodothyronine Deiodination in Humans

Mutations in the monocarboxylate transporter 8 (MCT8) gene, located on the X chromosome, were first reported in 2004 [8, 9]. Subsequently, 26 families, comprising more than 100 affected subjects have been identified (pers. observations). The phenotype has a thyroid and a neuropsychiatric component [see chapter by Grüters, pp. 118–126]. The thyroid phenotype consists of TFT abnormalities that are the opposite of those found in patients with SBP2 defects, namely high T_3, low T_4, low rT_3 but also slightly elevated TSH. These TFT abnormalities are found in males and to a lesser degree in carrier females.

The neuropsychiatric component consisting of severe motor and developmental delay, gait disturbance, dystonia, and poor head control is found only in males. Understanding of the mechanism mediating the TFT abnormalities and the involvement of TH metabolism was derived from investigations in vitro, using the patients' fibroblasts and in vivo, using a genetically engineered mouse, deficient in Mct8 [15]. Cultured skin fibroblasts from the patients have increased D2 activity and mRNA when cultured in the same hormonal milieu as fibroblasts from normal individuals. In Mct8KO mice, which replicate the characteristic thyroid phenotype observed in humans, the defect in cell uptake of TH is not uniform in all tissues. Due to redundancy in transmembrane transporters, T_3 uptake in liver is not impaired while that in brain is substantially reduced. The consequence of these opposite states of intracellular TH availability is an increase in brain D2 and liver D1, adding to the consumptive effect on T_4 levels and resulting in increased T_3 generation.

Acquired Deiodinase Defects in Humans

Until recently, known defects in TH metabolism in humans were all acquired. A frequently encountered one is the 'low T_3' syndrome in nonthyroidal illness [16]. A decreased serum T_3 level is the most common thyroid function abnormality in patients with acute illness and can be detected within 2 h after the onset of severe physical stress. As the severity of the illness progresses, there is a gradual development of a more complex syndrome associated with low levels of T_3 and T_4. Altered TH levels have been reported in starvation, acute and chronic medical illnesses, bone marrow transplantation, surgery, trauma, and can be seen in any severe systemic illness. Decreased 5′-monodeiodination reducing both the conversion of T_4 to T_3 and the degradation of rT_3 is the principal mechanism responsible for the decrease in circulating T_3 and increase in rT_3 levels in severe illness. With more prolonged illness, increased turnover of T_3 and T_4 and an alteration in thyrotropin (TSH) secretion play secondary roles. Some inflammatory cytokines such as TNF-α, IL-1 and IL-6 have been recently implicated at both central and peripheral levels. Exogenous administration of TNF-α and IL-1 in humans and animals replicates the TFT changes reported in the syndrome [16]. Certain pharmacologic agents (dopamine, amiodarone, corticosteroids) may also alter the pattern of TFTs in a similar way and this should be taken into consideration when evaluating patients with nonthyroidal illness and multiple tests at different time points are recommended.

Another form of acquired abnormality in TH metabolism is that caused by increased D3 content in hemangiomas [17]. The phenotype resolves with tumor involution or resection. Nine cases in infants, and one in an adult, have been reported. The phenotype is that of consumptive hypothyroidism, with increased TSH and marked elevation of rT_3 in the context of normal T_3 and T_4. Hemangiomas are the most common tumors in infancy, with a prevalence of 5–10% among 1-year-olds. The high level of D3 produced by the tumor inactivates T_4 by conversion to rT_3 at rates that exceed the synthetic capacity of the thyroid gland. D3 activity of the tumor was found to be 3- to 7-fold higher than that of term placenta, the human tissue with the highest D3 activity.

Mouse Models of Deiodinase Deficiencies

D1 has been recently inactivated in mice [18]. The general health and reproductive capacity of the D1KO mouse were seemingly unimpaired. Serum levels of T_4 and rT_3 were elevated, whereas those of TSH and T_3 were unchanged. However, D1 deficiency resulted in marked changes in TH metabolism and fecal excretion of endogenous iodothyronines was greatly increased. Although D1 is of questionable importance for the well-being of the euthyroid mouse, it may play a major role in limiting the detrimental effects of conditions that alter normal thyroid function, including hyperthyroidism and iodine deficiency.

Another model for D1 deficiency is a naturally occurring mouse strain with marked decrease in D1 activity. C3H mice, compared to those of the C57BL6 strain, have a 5- to 10-fold lower D1 activity [19]. Sequence comparison revealed that the Dio1 gene of C3H mice has a 21-nucleotide insertion in the promoter, containing 5 CTG repeats, and a 150-nucleotide expansion of highly repetitive sequences in intron 2. This was associated with a decrease in the Dio1 mRNA level, caused by a decreased rate of transcription. The thyroid phenotype was that of increased T_4 and rT_3 with normal T_3. Severe Se deficiency caused a similar pattern of TFTs in rats through reduction in D1 activity.

D2 was the first deiodinase to be deleted in the mouse through homologous recombination [20]. Except for minimal (9%) growth retardation in males, no gross abnormalities were observed in D2KO mice. No D2 activity has been observed in these animals under basal conditions or under stimuli, such as cold exposure or hypothyroidism. D2KO mice have defective auditory function, retarded differentiation of the cochlear inner sulcus and sensory epithelium, and deformity of the tectorial membrane [21]. This suggests that D2 is essential for hearing, and this hormone-activating enzyme confers on the cochlea the ability to generate its own T_3 at a critical developmental period. Overall development and reproductive function appeared normal with normal serum levels of T_3. However, T_4 and TSH levels were significantly elevated, 40 and 100%, respectively. This indicates that the pituitary gland of D2KO mice is resistant to the feedback effect of plasma T_4, caused by the reduction in the generation of intracellular T_3 through 5′-deiodination by D2. In fact, the elevated serum TSH levels were corrected by the administration of T_3.

Mice with targeted disruption of the *Dio2* gene were backcrossed into C3H strain with genetically low D1 expression to create the C3H-D2KO mouse [22]. Remarkably, these mice maintain normal serum T_3 levels and no decrease in expression of hepatic T_3-responsive genes. However, serum T_4 concentration is increased 1.2-fold relative to the already elevated level in C3H mice, and serum TSH is increased 1.4-fold. There is no further increase in rT3 in C3H-D2KO mice compared to the already 2.5-fold higher rT3 in C3H mice. C3H-D2KO mice have residual D1 activity in liver and kidney. A comparison of the C3H-D2KO animal model with mice with no D1 and D2 will provide further insight into the role of this small residual D1 present in the C3H-D2KO in maintaining normal serum T_3 levels.

D3 has been recently inactivated in mice, by replacing the Sec codon with cysteine, thus losing the characteristic enzymatic activity [23]. Mice heterozygous for the mutation showed decreased D3 activity when the null allele was inherited from the father and no change when it was inherited from the mother. This is due to imprinting, with the paternal Dio3 being predominantly expressed. Dio3 is highly expressed during fetal life and in the placenta which appears to be the basis for the occurrence of partial neonatal mortality and growth retardation in D3-deficient mice.

Mouse Models of Sec Incorporation Defects

Since the report in 1973 that GPx is a selenoprotein, more than 20 other Sec-containing proteins have been identified. Although novel selenoprotein genes have been identified in the human genome using bioinformatic approaches, their functions remain largely unknown. Selenoproteins with known function play critical roles in a variety of biological processes, several being involved in antioxidant defense [5, 24]. Their synthesis in vivo is highly selenium-dependent, and a hierarchy of selenoprotein expression is believed to exist when selenium is limiting. Sec incorporation has been initially studied in prokaryotes, and subsequently the components of the eukaryotic machinery have also been identified [13].

Knockout mice for the gene encoding $tRNA^{Sec}$ (Trsp) die early in embryonic development [25]. Different conditional knockout and transgenic lines have been created to overcome this early lethality in order to study the implications for the Sec incorporation and selenoprotein function [26–29]. There are 2 major isoforms of mammalian $tRNA^{Sec}$ that differ by a methyl group on U^{34}, the first nucleotide of the anticodon. These isoforms are utilized differently in selenoprotein biosynthesis, demonstrating a unique manner in which the Sec machinery has evolved to express this class of proteins under conditions of selenium deprivation. There is no known mouse model of a SBP2 defect and it is possible that a complete deficiency might be lethal in embryonic stem cells.

Conclusions and Speculations

The phenotype produced by *SBP2* mutations is characterized by low T_3, high T_4 and high rT_3 and variable growth delay without other obvious abnormalities resulting from a general deficiency in selenoprotein synthesis [7].

SBP2 is believed to be the major determinant of Sec incorporation as its in vitro addition increases selenoprotein synthesis by 20-fold, whereas its immunodepletion eliminates Sec incorporation [13]. The absence of more prominent and generalized symptoms in the patients described above is undoubtedly due to the partial loss of SBP2 function.

It is remarkable that affected individuals have a similar thyroid phenotype as mice with combined complete D1 and D2 targeted disruption, without having a defect in these loci. As the subjects described herein, these mice have high serum T_4 and TSH concentration, low T_3 and markedly elevated rT_3 [V. A. Galton, pers. commun]. In contrast to D2KO mice, the affected children had normal hearing, probably due to partial deiodinase deficiency.

The global effect of SBP2 deficiency on the synthesis of selenoproteins has been documented. Although the reduction in GPx and SePP is not trivial, thyroid abnormalities resulting from decreased D2 activity and likely also D1 and D3 appear to dominate the clinical phenotype. Among the known selenoproteins, the UGA codon of D2 is most distant from the SECIS element and the half-life of the protein is less than 45 min. These factors and the hierarchy among selenoproteins might aggravate a deficit in Sec incorporation in D2, producing this specific thyroid phenotype.

The connection between Se and fertility is known as is the role of selenoproteins in sperm maturation [28]. The propositus of the Saudi family had delayed puberty, but we do not know if this is a characteristic of the syndrome or a coincidental occurrence, as the other 3 patients are below 9 years of age. In various mouse models of deficiencies of selenoproteins, such as GPx1 and GPx2 KOs, it was shown that exposure to oxidative stress, UV irradiation, administration of neurotoxic agents and exposure to viruses that infect the digestive tract has more deleterious effects compared to their normal littermates. These result in skin cancer, neuronal damage, and gastric cancer, respectively [30].

Finally, the consequences of the observed SBP2 defects might be underestimated in these young subjects that, with age, could present additional manifestations, such as decreased fertility [28] and propensity to develop cancer due to impaired antioxidative protection [30]. Having this Sec incorporation defect and the low levels of serum Se, it is possible that Se supplementation would be beneficial to these patients. The finding of a predominant thyroid phenotype in SBP2 defects highlights the importance of selenoproteins for thyroid feedback regulation. The identification of additional patients and their long-term follow-up are important in further characterizing this recently described defect.

Acknowledgment

This study was supported by grants DK17050, DK20595 and RR00055 from the National Institutes of Health (S.R.) and Howard Hughes Medical Institute Predoctoral Fellowship (A.M.D.).

References

1 Friesema EC, Jansen J, Milici C, Visser TJ: Thyroid hormone transporters. Vitam Horm 2005;70: 137–167.
2 Bianco AC, Salvatore D, Gereben B, Berry MJ, Larsen PR: Biochemistry, cellular and molecular biology, and physiological roles of the iodothyronine selenodeiodinases. Endocr Rev 2002;23: 38–89.
3 Refetoff S, Dumont JE, Vassart G: Thyroid disorders; in Scriver CR (ed): The Metabolic and Molecular Bases of Inherited Diseases. New York, McGraw Hill, 2001, chap 158.
4 Driscoll DM, Copeland PR: Mechanism and regulation of selenoprotein synthesis. Annu Rev Nutr 2003;23:17–40.
5 Hatfield DL, Gladyshev VN: How selenium has altered our understanding of the genetic code. Mol Cell Biol 2002;22:3565–3576.
6 Copeland PR, Fletcher JE, Carlson BA, Hatfield DL, Driscoll DM: A novel RNA binding protein, SBP2, is required for the translation of mammalian selenoprotein mRNAs. EMBO J 2000;19: 306–314.
7 Dumitrescu AM, Liao XH, Abdullah MS, et al: Mutations in SECISBP2 result in abnormal thyroid hormone metabolism. Nat Genet 2005;37:1247–1252.
8 Dumitrescu AM, Liao XH, Best TB, Brockmann K, Refetoff S: A novel syndrome combining thyroid and neurological abnormalities is associated with mutations in a monocarboxylate transporter gene. Am J Hum Genet 2004;74:168–175.
9 Friesema EC, Grueters A, Biebermann H, et al: Association between mutations in a thyroid hormone transporter and severe X-linked psychomotor retardation. Lancet 2004;364:1435–1437.
10 Botero D, Gereben B, Goncalves C, et al: Ubc6p and ubc7p are required for normal and substrate-induced endoplasmic reticulum-associated degradation of the human selenoprotein type 2 iodothyronine monodeiodinase. Mol Endocrinol 2002;16:1999–2007.
11 Curcio-Morelli C, Zavacki AM, Christofollete M, et al: Deubiquitination of type 2 iodothyronine deiodinase by von Hippel-Lindau protein-interacting deubiquitinating enzymes regulates thyroid hormone activation. J Clin Invest 2003;112:189–196.
12 Lescure A, Allmang C, Yamada K, Carbon P, Krol A: cDNA cloning, expression pattern and RNA binding analysis of human selenocysteine insertion sequence (SECIS) binding protein 2. Gene 2002;291:279–285.

13 Copeland PR: Regulation of gene expression by stop codon recoding: selenocysteine. Gene 2003;312:17–25.
14 Kohrle J: Selenium and the control of thyroid hormone metabolism. Thyroid 2005;15:841–853.
15 Dumitrescu AM, Liao XH, Weiss RE, Millen K, Refetoff S: Tissue-specific thyroid hormone deprivation and excess in monocarboxylate transporter (mct) 8-deficient mice. Endocrinology 2006;147:4036–4043.
16 Umpierrez GE: Euthyroid sick syndrome. South Med J 2002;95:506–513.
17 Huang SA, Tu HM, Harney JW, et al: Severe hypothyroidism caused by type 3 iodothyronine deiodinase in infantile hemangiomas. N Engl J Med 2000;343:185–189.
18 Schneider MJ, Fiering SN, Thai B, et al: Targeted disruption of the type 1 selenodeiodinase gene (Dio1) results in marked changes in thyroid hormone economy in mice. Endocrinology 2006;147: 580–589.
19 Maia AL, Berry MJ, Sabbag R, Harney JW, Larsen PR: Structural and functional differences in the dio1 gene in mice with inherited type 1 deiodinase deficiency. Mol Endocrinol 1995;9:969–980.
20 Schneider MJ, Fiering SN, Pallud SE, et al: Targeted disruption of the type 2 selenodeiodinase gene (DIO2) results in a phenotype of pituitary resistance to T4. Mol Endocrinol 2001;15: 2137–2148.
21 Ng L, Goodyear RJ, Woods CA, et al: Hearing loss and retarded cochlear development in mice lacking type 2 iodothyronine deiodinase. Proc Natl Acad Sci USA 2004;101:3474–3479.
22 Christoffolete MA, Arrojo EDR, Gazoni F, et al: Mice with impaired extrathyroidal thyroxine to 3,5,3′-triiodothyronine (T3) conversion maintain normal serum T3 concentrations. Endocrinology 2007;148:954–960.
23 Hernandez A, Fiering S, Martinez E, Galton VA, St Germain D: The gene locus encoding iodothyronine deiodinase type 3 (Dio3) is imprinted in the fetus and expresses antisense transcripts. Endocrinology 2002;143:4483–4486.
24 Schomburg L, Schweizer U, Kohrle J: Selenium and selenoproteins in mammals: extraordinary, essential, enigmatic. Cell Mol Life Sci 2004;61:1988–1995.
25 Bosl MR, Takaku K, Oshima M, Nishimura S, Taketo MM: Early embryonic lethality caused by targeted disruption of the mouse selenocysteine tRNA gene (Trsp). Proc Natl Acad Sci USA 1997;94:5531–5534.
26 Schweizer U, Schomburg L, Savaskan NE: The neurobiology of selenium: lessons from transgenic mice. J Nutr 2004;134:707–710.
27 Carlson BA, Novoselov SV, Kumaraswamy E, et al: Specific excision of the selenocysteine $tRNA^{Ser}Sec$ (Trsp) gene in mouse liver demonstrates an essential role of selenoproteins in liver function. J Biol Chem 2004;279:8011–8017.
28 Carlson BA, Xu XM, Gladyshev VN, Hatfield DL: Selective rescue of selenoprotein expression in mice lacking a highly specialized methyl group in selenocysteine tRNA. J Biol Chem 2005; 280:5542–5548.
29 Weiss Sachdev S, Sunde RA: Selenium regulation of transcript abundance and translational efficiency of glutathione peroxidase-1 and -4 in rat liver. Biochem J 2001;357:851–858.
30 Chu FF, Esworthy RS, Chu PG, et al: Bacteria-induced intestinal cancer in mice with disrupted Gpx1 and Gpx2 genes. Cancer Res 2004;64:962–968.

Prof. Samuel Refetoff
University of Chicago, MC 3090
5841 S. Maryland Ave.
Chicago, IL 60637 (USA)
Tel. +1 773 702 6939, Fax +1 773 702 6940, E-Mail refetoff@uchicago.edu

Van Vliet G, Polak M (eds): Thyroid Gland Development and Function.
Endocr Dev. Basel, Karger, 2007, vol 10, pp 140–172

Papillary and Follicular Thyroid Cancers in Children

Vasyl Vasko[a], *Andrew J. Bauer*[a,b], *R. Michael Tuttle*[c], *Gary L. Francis*[d]

[a]Department of Pediatrics, Uniformed Services University of the Health Sciences, Bethesda, Md., [b]Department of Pediatrics, Walter Reed Army Medical Center, Washington, D.C., [c]Department of Endocrinology, Memorial Sloan Kettering Cancer Center, New York, N.Y., and [d]Department of Pediatrics, Medical College of Virginia, Virginia Commonwealth University, Richmond, Va., USA

Abstract

Benign and malignant neoplasms of the thyroid are uncommon during childhood, but they create diagnostic problems for the clinician to identify malignant legions that must be removed as well as medically important lesions that require treatment. Throughout the 20th century there was a rapid increase in the incidence of thyroid neoplasms which we now know were induced by ionizing radiation acquired from radiation therapy for benign medical conditions or from environmental sources such as nuclear testing and accidents [Cancer 1961;14:734–743]. Over recent decades, there have been major advances in our understanding of the molecular biology and clinical management of thyroid neoplasms. We hope that the reader of this chapter will find this information of benefit in the clinical management of children with thyroid neoplasms and will be encouraged to study remaining controversial issues. We have divided this chapter into two major sections, the first of which pertains to thyroid nodules and the second to well differentiated thyroid cancers including papillary, follicular and other variants.

Thyroid Nodules in Children

The incidence and prevalence of thyroid nodules during childhood are not well known. Estimates suggest that the incidence is about 1–1.5% [1, 2]. However most clinicians do not identify nodular disease of thyroid in such a high proportion of children. It has also been estimated from postmortem and ultrasound (US) studies that 13% of young adults have thyroid nodules and that 50% of adults will develop thyroid nodules by the age of 50 years [2]. From these data it would appear that increasing age is a risk factor for nodular disease of the thyroid. It is well known that iodine deficiency is another risk factor [3]. Whether or not these associations relate to increasing thyrotropin (TSH) levels throughout life, an increased prevalence of iodine deficiency in older patients, or cumulative exposure to environmental toxins is not at all clear.

Regardless of the cause, we now know that many thyroid nodules contain mutations in either the stimulatory subunit of the guanine nucleotide triphosphate-binding protein (Gsα) or a genetic recombination between the paired homeobox gene (Pax 8) and the peroxisome proliferator gene (PPAR-γ) [3]. Both of these mutations have been directly linked to the induction and growth of follicular neoplasms in adults.

One of the most important distinctions between nodular disease of the thyroid during childhood and adult life is the probability that a thyroid nodule will be malignant. In adults the proportion of thyroid nodules that are proven to be malignant is relatively low, on the order of 5–10% [4]. However, during childhood 30–50% of thyroid nodules are malignant [5]. For this reason, it is of critical importance to distinguish benign from malignant thyroid nodules in children. The diagnostic tools available to accomplish this include thyroid scintigraphy, thyroid US, and fine needle aspiration (FNA). A recent study by Vierhapper et al. [6] showed that routine determination of serum calcitonin levels in patients with thyroid nodules facilitates early diagnosis and treatment of medullary thyroid carcinoma (MTC). The authors suggested that routine serum calcitonin levels should be determined in all patients; however, the benefit of this approach has not yet been specifically demonstrated for children.

Thyroid scintigraphy is not particularly beneficial in the distinction of benign from malignant lesions [7]. Based on scintigraphy, thyroid nodules are classified as either 'cold' when the tracer uptake is less than that of the surrounding normal thyroid, 'warm' when uptake is seen in both the nodule and the surrounding tissue, and 'hot' when the nodule shows avid tracer uptake and the surrounding thyroid is fully suppressed. Only in the latter case, that of a 'hot' nodule, are the risks of malignant disease reduced. Despite the reassuring finding of a 'hot' nodule, malignant thyroid neoplasms with 'hot' uptake on scintigraphy have been reported and have been linked to the presence of activating mutations in the Gsα or the TSH receptor [8].

The increasing use of thyroid US has allowed several characteristics of benign and malignant lesions to be identified. These are not used as the sole criteria on which to base clinical management, but are used as additional factors to support or refute surgery [9]. A translucent halo surrounding the lesion, homogenous echotexture, and lack of internal calcification are more commonly associated with benign lesions [10]. In contrast, indistinct margins, internal calcifications particularly microcalcifications (indicative of psammoma bodies) and variable echodensity are more commonly associated with malignant lesions [10].

Blood flow to the nodule has been assessed with the newer technique of power Doppler. DeNicola et al. [11] examined 86 thyroid nodules for flow pattern and resistive index (RI). The flow pattern was ranked on a scale of 0–4, and the RI was defined by the average of 1–3 values. The average RI in nonneoplastic nodules was 0.588 compared to 0.763 in malignant nodules. Peripheral flow or predominantly peripheral flow was seen in 93.5% of nonneoplastic nodules but also 20% of malignant nodules. Predominantly central flow was seen in 70% of malignant nodules, but also in 28.6% of adenomas. Despite a statistically significant correlation between central blood flow and malignant disease, power Doppler could not be used to exclude malignant disease because 20% of malignant nodules had predominantly peripheral flow.

Perhaps the greatest strengths of US lie in the exquisite sensitivity of this technique to identify thyroid nodules, to ascertain additional lesions that were not evident on palpation of the gland, and to follow the size and characteristics of small lesions over time [12].

FNA has been widely employed in the study of adult neoplasms to distinguish benign from malignant disorders. Reviews of multiple series such as those of Mazzaferri [13, 14] (9,119 FNAs) and Gharib et al. [15, 16] (18,183 FNAs) confirm the overall value of FNA with sensitivity ranging from 68 to 98% and specificity from 72 to 100%. The major limitations of FNA are related to the skill of the operator and the experience of the cytopathologist [17–19].

Due to the uncommon finding of thyroid nodules in children, not many centers have sufficient experience in this age group to comment on the sensitivity and specificity of FNA specifically for children. However, Gharib et al. [15, 16] performed 10,971 FNAs, of which 57 were performed on patients younger than 17 years of age. Sixty-six percent of the FNAs were benign, 15% were malignant, 6% were suspicious and 13% were nondiagnostic. There were no false-positive results but there was one false-negative result (1/7 malignant lesions, 14%). Corrias et al. [9] used FNA along with several clinical features in the examination of 41 children with thyroid nodules. There were no false-negative results but all patients were at low risk of malignant disease. Smaller series by Degnan et al. [20] and Raab et al. [21] also report individual false-negative FNA leading some authors to suggest that all thyroid nodules in children should be surgically removed.

In general, FNA is most valuable for identifying papillary thyroid cancer (PTC), which can be distinguished by the nuclear characteristics on cytology. Cells may be arranged in clusters with occasional papillary fragments. Individual cells have enlarged nuclei, dense chromatin, and variable shape [22]. Nuclear inclusions and grooves are rarely seen if the preparation was stained with Diff-Quik, but are common with the Papanicolaou stain [22]. Psammoma bodies may also be seen.

Unfortunately tissue architecture cannot be ascertained from FNA and for that reason it is impossible to distinguish benign follicular neoplasms from malignant follicular thyroid cancer (FTC) [23]. The latter diagnosis is based on the presence of vascular or capsular invasion. Results of FNA are therefore classified as (1) consistent with PTC, (2) suspicious for malignancy, (3) consistent with a follicular neoplasm, (4) consistent with a benign lesion, or (5) inadequate/nondiagnostic findings. Cytology consistent with PTC, suspicious for malignancy and all follicular neoplasms are removed surgically [18, 19].

An area of controversy is the management of the patient with an inadequate or nondiagnostic FNA. Orija et al. [24] surveyed 143 physicians who perform FNA. Most of them (57.5%) had fewer than 10% nondiagnostic FNA. The survey results showed that the vast majority (82%) would repeat the FNA, while 17% would monitor the size of the nodule and 1% would refer for surgery. No one was willing to repeat the FNA more than 3 times. Based on the higher probability of malignant disease in children with thyroid nodules, it is not clear if this approach would be recommended by a majority of practitioners. However, it is our personal opinion that nondiagnostic findings on FNA in children warrant either repeat FNA for adolescent low-risk patients or surgical removal for younger children and for high-risk lesions.

Recent attempts to improve diagnosis from nondiagnostic specimens have focused on the potential utility of molecular markers whose expression is upregulated in thyroid cancer. Haugen et al. [25] examined telomerase activity in 24 thyroid tumors using a telomeric repeat amplification protocol (TRAP). TRAP activity was detected in 11 of 20 thyroid carcinomas (55%) but was not detected in 4 benign adenomas, 3 FTC, 1 MTC or 1 anaplastic thyroid carcinoma. Six of 7 invasive PTC (86%) had TRAP activity.

Based on these initial observations, Lerma and Mora [26] studied telomerase activity in patients with nonconclusive cytology on FNA. Telomerase activity was detected in 6 of 18 thyroid neoplasms including 1 of 3 follicular adenomas (FA), 3 of 11 PTC, 0 of 1 FTC, 1 of 2 MTC, and 1 of 1 lymphoma. Detection of telomerase activity helped to confirm neoplasia in 6 of 23 (26%) suspicious nodules.

Telomerase analyses have also been used to attempt to distinguish FTC from benign FA. Kammori et al. [27] used TRAP assays of telomerase activity and in situ hybridization to detect telomerase gene expression in paired FNA and tissue samples from 6 FTC and 15 FA. Telomerase activity was detected in all 6 FTC and 5 of the 15 (33%) FA. Telomerase gene expression was detected in all of the FTC and 1 of the adenomas (7%) and in 4 of the 6 (67%) FTC biopsy specimens obtained using FNA. Similar studies have not been performed in childhood thyroid cancers; however, Straight et al. [28] did detect telomerase expression in a number of childhood thyroid cancers where expression was associated with a more aggressive clinical course. Future study will be required to determine the utility of telomerase assays in the detection of malignant lesions by FNA in this age group.

Cystic lesions also present a diagnostic challenge for FNA as cysts may harbor occult PTC. Immunostaining for the tumor marker, galectin-3, has been used to improve the detection of occult PTC by FNA of cystic lesions. Papotti et al. [29] examined samples from 32 cystic PTC and 12 benign cysts obtained by FNA. Almost all cystic PTC (29/30, 97%) were galectin-3 positive. In comparison, cytology made a correct diagnosis in only 25 of 32 (78%). All benign cysts were negative for galectin-3.

Another controversial area is the follow-up of patients with benign lesions on FNA, specifically regarding the value or requirement of repeating the FNA over time. Orlandi et al. [30] performed annual FNA on 306 patients (aged 14–84 years) with benign initial FNA over 2–12 years. Of all patients, 97.7% continued to have benign cytology, while only 3 patients (0.98%) developed suspicious findings and 4 (1.30%) developed PTC. Of the 4 with PTC, 1 was diagnosed on the second and 3 on the third repeat FNA. The authors suggest that at least three annual FNA should be performed in order to reduce the risk of missing a PTC. Despite these suggestions, most adult endocrinologists perform repeat FNA only for large nodules, nodules that increase in size, or nodules that develop suspicious US findings on serial examination. Although some patients in the study by Orlandi et al. were adolescents (14 years of age and older) it is not clear if the data pertain to younger children (less than 10 years of age) who are generally believed to have a higher risk of malignant disease [31].

Many clinicians prescribe thyroid hormone to reduce the size of benign nodules. However, Gharib [32], Gharib and Mazzaferri [33] and Giuffrida and Gharib [34] showed that only 10–20% of benign lesions in adults respond to thyroid hormone suppression (defined as a decrease of ≥50% in size). A similar response rate was also seen in the placebo group. A recent meta-analysis by Sdano et al. [35] reported outcomes for 609 subjects with thyroid nodules. Patients treated with thyroid hormone suppression were almost twice as likely to show reduction in nodule volume. Unfortunately, 8 patients would be exposed to the risks of thyroid hormone suppression for every 1 patient who benefits. Papini et al. [36] showed that thyroid hormone suppression reduced the risk of additional nodule formation from 28.5 to 7.5% over 5 years. The risks of thyroid hormone suppression are not well known for children and adolescents and the data from these studies appear to favor thyroid hormone suppression with goals of reducing nodule volume and the risk of additional nodule formation. For these reasons, many clinicians including the authors of this chapter, prescribe thyroid hormone suppression for children with benign thyroid nodules. Optimal TSH suppression is controversial since most benign nodules remained stable and benign over long periods of time and since suppression of TSH to levels below the normal range (<0.3 mIU/ml in most assays) induces subclinical hyperthyroidism [37–39].

We use several clinical features in our decision of whether to follow or remove an apparent benign nodule. These include (1) patient age, (2) past history of ionizing radiation, (3) family history of thyroid nodules or thyroid cancer, (4) US characteristics, and (5) FNA cytology [40]. We believe the probability of malignant disease is greater in children under 10 years of age and are less inclined to follow benign-appearing lesions in this age group. However, for low-risk patients such as the adolescent with no history of ionizing radiation exposure, benign US features and benign FNA cytology, we have been willing to follow these lesions over time with serial US examination. Nodules that increase in size are re-evaluated with FNA or surgically removed [32–34].

Differentiated Thyroid Carcinoma

Incidence and Radiation-Associated Risks

The first report of differentiated thyroid carcinoma in a child was that of Ehrhardt [41] in 1902. During the 1950s there was a dramatic increase in the incidence of childhood PTC that was directly linked to the use of external beam radiation therapy for the treatment of benign medical conditions such as acne and tinea capitis [42]. A landmark article published by Winship and Rosvall [1] in 1961 clearly made the association between radiation exposure and the subsequent development of PTC often 1–2 decades later. As a result, ionizing radiation therapy was discontinued as a treatment for these disorders and the incidence of PTC declined. However, in recent decades the incidence of PTC has increased worldwide perhaps as a result of contamination from the Chernobyl nuclear accident [43]. Not only the incidence, but also the mortality for thyroid cancers has increased over the last 20 years. Currently, differentiated thyroid carcinomas including papillary (PTC) and follicular (FTC) variants account for 1–2% of all childhood malignant disease [44]. The annual incidence is approximately 1/1,000,000 prior to puberty but this increases to approximately 30/1,000,000 in girls and 6/1,000,000 in boys during puberty [44]. Reasons for the greater incidence in young women are not yet clear.

The powerful relationship between ionizing radiation and the development of PTC has been a subject of great interest. Nikiforov [45] recently identified a unique geometric packaging of chromosomes within the nucleus that predisposes to rearrangements between the ret proto-oncogene and other genes that result in unregulated ret transcription. Buckwalter et al. [46] have shown that these rearranged ret proto-oncogenes (ret/PTC) are sufficient to induce thyroid cancers in transgenic mice.

Nikiforov [45] used two-color fluorescent in situ hybridization and three-dimensional microscopy to map the locations in interphase nuclei of the ret proto-oncogene and the other genes involved in ret/PTC recombinations. They found a juxtaposition of the RET proto-oncogene and H4 that could facilitate generation of ret/PTC-1 in 35% of normal human thyroid cells. They also found that the RET/PTC-3 rearrangement, which is formed by fusion of the ELE1 and RET genes, is highly prevalent in radiation-induced post-Chernobyl PTC [47]. They found juxtaposition and opposite alignment of ELE1 and RET in each tumor suggesting that a single ionizing event could energize both genes and allow for a recombinant event between the RET proto-oncogene and ELE1 [47]. These elegant studies provide the first molecular insights into the mechanisms by which ionizing radiation could specifically induce thyroid cancer.

Studies by Ron et al. [48] showed that the risk of developing thyroid cancer is related to the absorbed dose of radiation with a relative risk of 7.7 per absorbed gray and studies by Shore [49] showed that children are at least 10-fold more sensitive than adults to the effects of ionizing radiation on the thyroid. Of major interest from a public health perspective were differences in the management and outcome of children exposed to radiation following the Chernobyl nuclear accident. Children in Belarus and Ukraine received no prophylaxis and had almost a 200-fold increase in the incidence of PTC [50]. In contrast, children in Poland received stable iodine at doses of 15 mg for newborns, 50 mg for children less than 5 years of age, and 70 mg to all others [51]. Overall, 10.5 million doses were administered to children and 90% of all children in Poland received at least 1 dose of stable iodide. Nauman and Wolff [51] reviewed the results of this widespread prophylaxis that revealed only minor complications. A few (0.37%) newborns had a transient increase in TSH and a concomitant reduction in thyroxine (T_4) levels. There were no long-term effects on thyroid function and 91.4% of the population had normal TSH values. Most importantly, there was no increase in the incidence of thyroid nodules, thyroid cancer, goiter, or hypothyroidism.

Presentation

The most common presentation of thyroid carcinoma in children is that of a thyroid nodule [52]. As discussed in the preceding section, thyroid nodules are uncommon in children (1–1.5%) but when present, they are frequently malignant (30–50%) [53]. Children are also more likely than adults to present with widely disseminated disease at diagnosis. Approximately 80% of children with PTC have disease involving the cervical lymph nodes, 20% have extrathyroidal extension, and distant pulmonary metastasis is present in 5–6% [52]. If these were adults, mortality would be very high. Tubiana et al. [54] showed that mortality for adult thyroid cancer increases with age from under 1% for adults less than 50 years of age to almost 50% for adults greater than 70 years of age. Likewise mortality increased with tumor size, from less than 1% in those with small tumors (less than 2 cm in diameter) to 50% for those with tumors greater than 7 cm. Finally, mortality increased in adults with distant metastasis reaching approximately 70% for those with pulmonary metastasis and 40% for those with recurrent disease.

In contrast, however, children with thyroid cancer rarely die of disease. Studies by Powers et al. [55, 56] showed that mortality was less than 15% in all series of childhood thyroid cancers and in most series was as low as 1–2%. Even after recurrence, none of the 7 children in their study died of disease, compared to 38.8% mortality in adults. They also found that mortality was lower in children with distant metastasis than in adults.

Brink et al. [57] followed 14 children for up to 45 years, during which time, none died even following recurrence. However, 7 of these children developed persistent disease. Similar data have been published by LaQuaglia et al. [58]. They followed 83 children with distant metastasis for an average of 10 years. Overall survival was 100% and progression-free survival was 65–70% over 5 years. Again, this study found that 30–45% of children with pulmonary metastasis develop stable but persistent disease. This propensity for PTC to develop stable, persistent disease in children with pulmonary metastasis is another of the striking differences between PTC in children and adults.

In an attempt to determine if these differences could be related to a different molecular signature, we and others have investigated mutation profiles in pediatric and adult thyroid cancers. Mutations in the RAS proto-oncogene have been identified in 12% of adult PTC, 29% of FTC, and 50% of anaplastic thyroid carcinoma [59–61, this study]. However, Nikiforov et al. [62] found a low incidence of RAS mutations in a series of childhood thyroid cancers that arose following the Chernobyl incident. Fenton et al. [63] examined a series of spontaneous PTC in United States children and found only 2 PTC with RAS mutations for an overall prevalence of 6.5%.

The ret/PTC gene rearrangements are known to occur in adult PTC with an overall incidence ranging from 23 to 76% [64–67]. In spontaneous PTC these are predominantly ret/PTC1. However, in radiation-induced disease, ret/PTC3 is most common [47]. Fenton et al. [68] showed that the prevalence of ret/PTC rearrangements was much higher in childhood PTC, approximately 45%. However, there was no correlation between the presence of a ret/PTC rearrangement and patient age, tumor size, extent of disease, or the risk of recurrence.

More recently, BRAF mutations have been identified as the most common in adult PTC [69–71]. Of these, the T1796A transversion has been demonstrated in 42% of adult PTC [this study, 69, 72]. Penko et al. [73] amplified the T1796A transversion site in 13 spontaneous PTC from children and found that none of these tumors contained a T1796A transversion. The prevalence of the T1796A transversion was significantly less than that reported in adults (42%). Of interest, 58% of the tumors in her study contained a ret/PTC rearrangement and none contained a RAS mutation. The tumors generally exhibited low-risk features. The patients had a favorable age distribution (10–21 years) and the tumors were not particularly large (0.7–2.9 cm in diameter). At diagnosis, 7 of the PTC were DeGroot class 1 (confined to the gland), 5 were DeGroot class 2 (regional lymph node involvement), and 1 was DeGroot class 3 (extending beyond the thyroid). None of the tumors were DeGroot class 4 (distant or pulmonary metastases) and none recurred over a median follow-up of 66 months. These data suggest the possibility that absence of a BRAF mutation might be associated with more favorable outcome.

PTC and FTC from children are generally well-differentiated and one can detect the thyroid-specific nuclear transcription factor [thyroid transcription factor 1 (TTF-1)] by immunohistochemistry in 78% of thyroid cancers from children [74]. TTF-1 is critical for maintaining expression of TSH receptor, sodium iodide symporter (NIS) and thyroglobulin (Tg), all of which are markers of differentiated thyroid tumors [75, 76]. In addition, the paired homeobox gene, Pax 8, is also important for thyroid differentiation and can be demonstrated on immunohistochemistry in the nucleus of 36% of childhood PTC and in the cytoplasm of 85% of these tumors [77]. The NIS is another marker of well-differentiated tumors and is also important for treatment by promoting uptake of radioactive iodine into thyroid cancers. Patel et al. [78] detected the NIS by immunohistochemistry in 36% of childhood PTC. Despite the fact that NIS was diffuse and cystoplasmic in location, there were important correlates between the intensity of NIS staining and clinical outcome. Twelve tumors had detectable NIS expression and 20 did not. Recurrent disease developed only in those patients who had undetectable NIS (overall incidence of 20%). In addition, the patients with tumors in which NIS was detected were successfully treated with lower cumulative doses of radioactive iodine.

In children, thyroid tumors also induce an immune response characterized by lymphocytic infiltration and by the presence of proliferating lymphocytes. Gupta et al. [79] showed that the risk of recurrence was directly correlated with the absence of lymphocytic infiltration. None of the tumors with more than 10 proliferating lymphocytes per high power field developed recurrent disease. However, almost 60% of tumors from which proliferating lymphocytes were absent developed recurrent disease within the first 2 years. Modi et al. [80] also recently showed that these lymphocytes consist of a mixture of CD4-, CD8- and CD19-positive cells suggesting involvement of both humoral- and cellular-mediated immune responses. Of interest, tumors that contained a combination of CD4-, CD8-, and CD19-positive lymphocytes also contained a higher number of proliferating lymphocytes per high power field (44.6 ± 22.9/ high power field) when compared to tumors with any other combination of lymphocytes (3.4 ± 2.1 proliferating lymphocytes/high power field).

Treatment

The management of children with differentiated thyroid cancer has been controversial. Many factors contribute to this dilemma. First, the disease is uncommon. Second, life-long follow-up is required to demonstrate improvement in outcome as mortality generally occurs in the second to third decades after diagnosis. Finally, mortality rates are extremely low, on the order of 1%. For these reasons, prospective studies are unlikely to be conducted. All the data we currently possess regarding treatment of children with thyroid cancer are derived from retrospective analyses that are subject to selection bias, stratification of treatment plans based on the extent of disease at diagnosis, and short-term follow-up of 5–10 years in the majority of reports. Fortunately, the last few years have seen data beginning to appear from studies with 30–50 years of follow-up and this has allowed several important observations to be made regarding the success of our current treatment programs.

Hung and Sarlis [31] in a recent review conceptually stratify pediatric patients into two age groups, those less than 10 years of age with a high risk of recurrence and mortality, and those greater than 10 years of age in whom mortality and recurrence risks are lower and more similar to those seen in young adults. Most surgeons, according to Hung and Sarlis perform total or near-total thyroidectomy. This preference is based on several features of childhood thyroid cancer. First, 40% of all children with PTC have multifocal disease that could place them at higher risk of recurrence if less than total thyroidectomy is performed [52]. From molecular signatures we now know that these multifocal tumors represent individual clones and not intrathyroidal metastasis [81]. For that reason, all cells in the thyroid are at potential risk of malignant transformation. Second, many children diagnosed with PTC will have disseminated disease at the time of diagnosis and these patients will require radioactive iodine therapy [52]. This will be more effective in patients who have undergone total thyroidectomy prior to radioactive iodine ablation. Third, newer and more sensitive assays for serum Tg levels are now available and are used as a marker for persistent and/or recurrent disease [82]. However, Tg is produced in large quantities by any remaining normal thyroid tissue. For that reason Tg levels are useful as a tumor marker predominantly in patients who have undergone total thyroidectomy and radioactive iodine ablation [82].

Although retrospective studies have generally failed to document a survival advantage for patients managed with total thyroidectomy, the data do indicate a lower risk of recurrence for patients who undergo total or near-total thyroidectomy. Welch Dinauer et al. [52] published a series of 198 children with thyroid cancer spanning more than 4 decades. Eighty-one percent of the tumors were PTC and 19% were FTC. The presenting feature was a thyroid nodule in 84% of patients and palpable cervical lymph nodes in 23%. Almost 6% of the patients had pulmonary metastases at diagnosis and these patients tended to be younger (median age 10.5 years), to have larger tumors (4.5 cm in diameter), and were more likely to have multifocal disease (71.4%). The vast majority (82.5%) were treated with total or subtotal thyroidectomy. Forty-seven percent had cervical lymph node dissection and 58.4% had radioactive iodine ablation. Over time, 4.3% of patients developed recurrence. These also tended to be younger with a median age of 13.5 years.

When stratified for risk factors that predict recurrence at any location, tumor size >2 cm at diagnosis, the presence of metastatic disease at diagnosis, palpable cervical lymph nodes at diagnosis, and the presence of multifocal disease were individually predictive of recurrence.

Similar findings have also been reported by Borson-Chazot et al. [83]. They evaluated 74 patients less than 20 years of age. Forty-five percent of the patients had total or subtotal thyroidectomy. They stratified patients according to the presence or absence of palpable cervical lymph nodes at diagnosis, and confirmed that patients with palpable cervical lymph nodes at diagnosis were more likely to recur (53% vs. no recurrence in patients without palpable nodes) and to develop persistent disease (30% compared to none for those without palpable cervical nodes). Patients with palpable cervical nodes were also more likely to have diffuse sclerosing histology (63 compared to 4% for patients without nodal disease), and more likely to have multifocal disease (89 compared to 16% for those without nodal disease). They also had a higher incidence of pulmonary metastasis at diagnosis (20% compared to none for those without nodal disease). Of additional interest, several of the patients who did not have nodal disease were treated with lobectomy and 10% of them recurred in the contralateral lobe underscoring the importance of bilateral surgery for reduction in the risk of recurrence.

Ian Hay [84] evaluated 189 patients ranging in age from 3 to 20 years, 81% of whom had initial neck node involvement, and 5% had pulmonary metastasis at diagnosis. The risk of recurrence was reduced by more than 50% if a bilateral thyroid resection was performed compared to a unilateral lobe resection. Over 40 years of follow-up, the risk of recurrence following bilateral surgery was approximately 25%. However following lobectomy alone, the risk of recurrence was 60% ($p = 0.0049$). Similar data on the value of bilateral thyroid surgery were also published by Welch Dinauer et al. [85]. She examined only patients who were believed to have disease confined to the thyroid gland. In that group of low-risk patients, lobectomy alone was associated with a 57% recurrence risk corresponding to an 8.7-fold increase when compared to recurrence rates following subtotal or total thyroidectomy.

The value of lymph node dissection for the treatment of PTC in children has not been as well studied. However, Demidchek and Kontratovich [86] published follow-up data on 662 children who developed thyroid cancer after exposure to radiation from the Chernobyl nuclear accident. One hundred and ten of them (16.6%) required reintervention for disease. Seventy-five required reintervention for lymph node metastasis and 12 required reintervention for pulmonary metastasis. When the primary intervention included no lymph node dissection, 20.6% required secondary intervention. However, when bilateral lymph node dissection was performed, only 7.2% required secondary intervention.

The use of radioactive iodine ablation for the treatment of children with thyroid cancer continues to be debated. Hung and Sarlis [31] indicate that despite surgery, significant uptake (>0.3%) is usually found in the thyroid bed. For that reason, they suggest children should receive a 30-mCi ablative dose 6 weeks following initial surgery. This should be followed in 6 months by a thyroid hormone withdrawal preparation whole body scan and repeat radioactive iodine therapy if disease is detected. This should be repeated every 6 months until the serum thyroglobin level is <8 ng/ml and the whole body scan is negative. Until recently, it was difficult to document the efficacy of such an approach.

Welch Dinauer et al. [52] found that the risk of recurrence after surgery alone was 32% as compared to 34% following surgery plus radioactive iodine. It should be remembered, however, that 82.5% of the patients in their study were treated with total thyroidectomy and 58.4% received radioactive iodine ablation. Ian Hay [84] compared the outcomes for 92 patients who were treated by near-total or total thyroidectomy with that of 45 patients treated by near-total or total thyroidectomy plus radioactive iodine ablation. His data showed that the risk of recurrence was similar in both groups over a 40-year follow-up interval. However, both of these studies are retrospective and it is difficult to discern why individual patients were selected for radioactive iodine ablation. If the patients who received radioactive iodine had more extensive disease than patients who did not receive radioactive iodine, then similar recurrence rates would actually support the use of radioactive iodine.

A recent study by Chow et al. [87] strongly supports the use of radioactive iodine ablation in children with differentiated thyroid cancer. They treated 60 naïve and 14 recurrent patients who ranged in age from 8.6 to 20.9 years. They were followed for an average of 14 years. Eighty-two percent of the tumors were PTC and at diagnosis, 45% had cervical lymph node disease, 6.7% had pulmonary metastasis, and 20% of the PTC were multifocal. In contrast to other studies, the use of radioactive iodine ablation was standardized. Radioactive iodine was given to patients if tumor diameter was >1 cm, or cervical lymph node disease was present, or extrathyroidal extension was present, or residual postoperative disease remained in situ, or distant metastasis was present. They gave 80 mCi if no distant metastasis was identified and 150 mCi if distant metastasis was present. Over time, the recurrence rate for patients who did not receive radioactive iodine was 42%, but this was dramatically reduced to 6.3% in patients who received radioactive iodine ($p = 0.001$). Of additional importance, 20.8% of patients who did not receive radioactive iodine developed pulmonary metastasis. In contrast, none of the 32 patients who were given radioactive iodine developed pulmonary metastasis. However, the numbers of patients in both of these latter groups were too small to achieve statistical significance ($p = 0.1$).

Another potentially important observation from their study can be extrapolated from the management of children with small (<1 cm) PTC. Over time, 42% of the patients with these small PTC developed recurrent disease. This is another of the major differences between children and adults with PTC. Adults with small PTC are commonly managed as low-risk patients, but children appear to have a much higher recurrence risk from these small PTC.

When counseling families as to anticipated response to therapy, Powers et al. [56] showed that the extent of disease at diagnosis predicts the response to initial therapy. They reviewed outcomes for 47 patients, of whom 33 (70%) went into remission with initial therapy (total thyroidectomy, lymph node dissection, and radioactive iodine ablation). When stratified by DeGroot classification, 79% of class 1, 86% of class 2, and 100% of class 3 patients went into remission. However, none of the patients with DeGroot class 4 disease (pulmonary metastases) entered remission after initial therapy.

Powers et al. [55] also reported on the treatment of recurrent PTC in children. Although the numbers of patients were small, 4 children had recurrent disease in the thyroid bed, 1 in the local lymph nodes, 1 in the lungs, and 1 in the cervical lymph nodes plus bone. Overall 5 out of 6 patients (83%) achieved remissions that lasted from 10 to 99 months and none of the patients died of disease.

Selecting an optimal dose of radioactive iodine for thyroid remnant ablation or for treatment of regional or distant metastasis is very difficult. Hung and Sarlis [31] suggest that the doses of radioactive iodine should either be adjusted for body size or should be adjusted based on dosimetry. They would adjust the doses routinely given to adults (100–150 mCi for thyroid bed uptake, 150 mCi for cervical node disease, and 200 mCi for pulmonary uptake) on a body weight basis using a ratio between the child's weight and an estimated adult weight of

70 kg. Leboulleux et al. [88] suggested that the doses of radioactive iodine should approximate 1 mCi/kg body weight and should be repeated every 6 months until remission is achieved. In their experience, 80% of patients entered remission after 4–6 treatments. Reynolds [89] in treatment of thyroid cancer in childhood NIDDK, NIH suggested a dose of radioactive iodine of approximately 1.5 mCi/kg body weight.

Dosimetry might be useful in calculating doses of radioactive iodine, particularly for small children or for patients with persistent disease who have already received multiple treatments. Dosimetry might also be helpful in mitigating against the risks of second malignancy and pulmonary fibrosis. Benua and Leeper [90] reviewed the experience of the Memorial Sloan Kettering Cancer Treatment Center and showed that doses which provided less than 2 Gy of exposure to bone marrow and retention of <120 mCi at 48 h did not induce permanent bone marrow suppression. However, Dorn et al. [91] found that the bone marrow was the dose-limiting organ in 46% of cases while the lung was the dose-limiting organ in almost 10% of cases. We do not yet know the frequency with which the lung is actually the dose-limiting organ in children.

Occasional children have been treated using dosimetry-based radioactive iodine doses and yet these children developed pulmonary fibrosis (fig. 1). There could be several potential explanations for this. Computer-based programs for selecting the maximal-tolerable dose were derived from adult data. It is possible that these programs may need to be adjusted for children. Another potential explanation could be the production of growth factors by individual tumors that might induce pulmonary fibrosis. Fenton et al. [92] identified high levels of vascular endothelial growth factor (VEGF) expression in a group of pediatric thyroid cancers, and showed that recurrence was more common in those with the most intense VEGF expression. Lennard et al. [93] confirmed these findings in adult PTC and also found higher blood levels of VEGF in patients with the most extensive disease. Patel et al. [94] also found expression of the matrix metalloproteinases (MMPs) and MMP inhibitors in childhood thyroid cancers. Collectively these data support the hypothesis that some tumors express high levels of growth factors that might stimulate pulmonary fibrosis but a direct link between high levels of expression and pulmonary fibrosis has not yet been examined.

Prescription of thyroid hormone to suppress the serum TSH level is a critical element in all treatment protocols for differentiated thyroid cancer including those for children. DeGroot et al. [95] showed that TSH is a powerful growth stimulus for thyroid cancer and that thyroid hormone suppression reduces the risk of recurrence. Despite these important findings, the optimal level of TSH suppression has not yet been defined for children. Undersuppression would predispose to recurrent thyroid cancer while oversuppression would induce subclinical hyperthyroidism

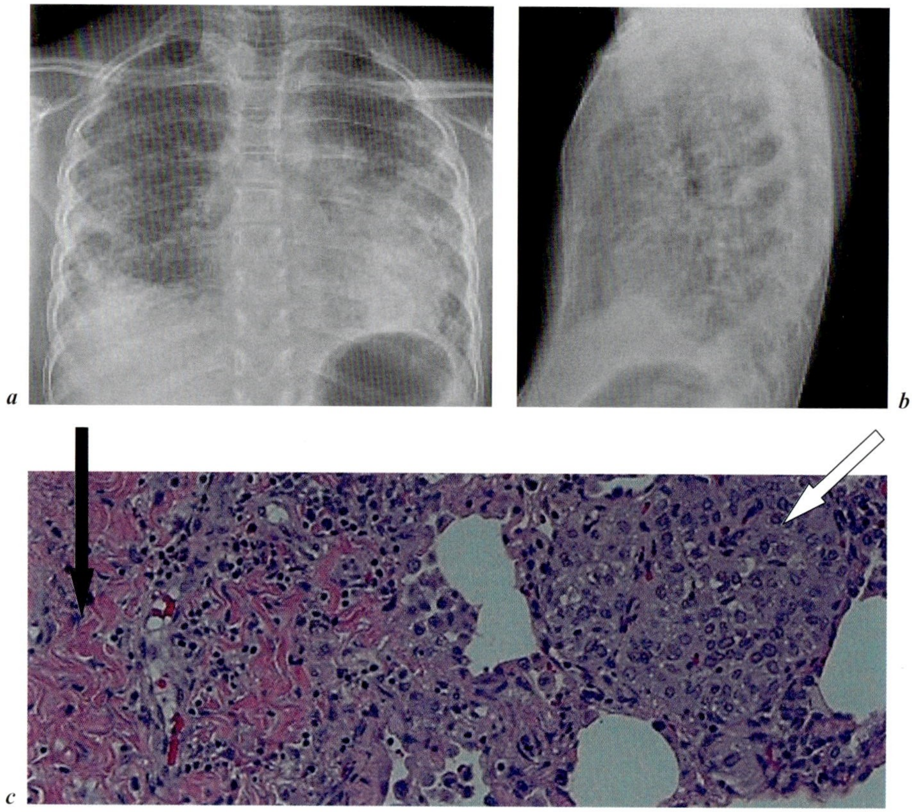

Fig. 1. ^{131}I-induced pulmonary fibrosis. Routine chest radiograph showing extensive pulmonary fibrosis (***a***, ***b***) and histological section from open lung biopsy showing extensive fibrosis (solid arrow) and metastatic papillary thyroid cancer (open arrow) (***c***).

that might have significant impact on growing children. Leboulleux et al. [88] recommended initial suppression of TSH to less than 0.1 μIU/ml followed by relaxation of TSH suppression to 0.5 μIU/ml once the patients enter remission. However, the long-term outcome of this approach has not been examined.

Many studies have now confirmed the extremely low mortality associated with this combined therapeutic approach (total thyroidectomy, lymph node dissection, radioactive iodine ablation and thyroid hormone suppression). Most demonstrate disease-specific mortality of 1% or less [96]. For these reasons, our expectation is that children with differentiated thyroid cancer will survive, but they may be at increased risk of complications from therapy.

Risks of Treatment

Extensive surgery (total or near-total thyroidectomy) carries increased risks when compared to less extensive procedures such as lobectomy. Although some have suggested that the risks of surgery are the same in children as in adults, many pediatric surgeons have not had such encouraging experience [88]. LaQuaglia and Telander [97] found that the risk for developing recurrent laryngeal nerve injury, hypoparathyroidism, infection, hematoma, seroma, hemorrhage, or hypertrophic scar approached 90% in the youngest patients (those <5 years of age). Other centers confirm these complications. Robie et al. [98] presented data from 126 children and adolescents, 17% of whom developed postoperative hypocalcemia and 3% developed recurrent laryngeal nerve injury. The probability of complications was significantly greater for those treated with total or subtotal thyroidectomy and approached 20%. None of the patients treated with lobectomy developed surgical complications. Patients with gross extension of disease beyond the thyroid had an even higher (40%) risk of complications. Kowalski et al. [99] reported their experience with 38 patients ranging in age from 4 to 18 years and followed for over 9.4 years. Surgical complications included transient hypocalcemia in 24%, permanent hypocalcemia in 16%, vocal cord paralysis in 5%, and wound infection in 5%. Van Santen et al. [100] presented a series of 26 children followed for up to 40 years. All were treated with total thyroidectomy and 60% received radioactive iodine ablation. Eighty-four percent had at least 1 adverse event. Thirty-one percent developed permanent hypoparathryoidism, 23% developed recurrent laryngeal nerve damage, and 8% developed Horner syndrome.

Radioactive iodine is associated with both short- and long-term risks. In the short term, many patients develop nausea and emesis. Antiemetics such as ondansetron hydrochloride are beneficial and fluid intake must be maintained to reduce radiation exposure to the bladder, colon and surrounding structures [40]. Patients can be encouraged to suck on sour hard candies (such as lemon drops) in an effort to stimulate salivary flow and reduce the risks of sialadenitis, but prospective trials of this approach have not confirmed benefit. Sialadenitis will generally develop within 2–4 days after radioactive iodine treatment, but can also occur up to 6 months after therapy. Most patients treated with over 100 mCi radioactive iodine develop hypogeusia, and some develop xerostomia [101]. Dorn et al. [91] showed that platelet counts reach a nadir of 50–100,000 and that white blood counts reach a nadir of 2,000–4,500 approximately 4–6 weeks after radioactive iodine therapy.

Over months to years, transient increases in follicle-stimulating hormone have been reported in adolescent males [102–104] while ovarian function tends to be preserved in women and pregnancy outcomes appear to be normal

[105–109]. However, a conservative recommendation is to avoid pregnancy for at least 6–12 months after treatment [110].

Of more serious concern are long-term follow-up data presented by Ian Hay [84]. He noted increased overall mortality in children who had thyroid cancer when compared to age-matched controls. Only 60% of childhood thyroid cancer victims were alive at the age of 60 years. In contrast, 75% of the control population was alive at the age of 60 years ($p = 0.001$). This excess mortality correlated with use of either radioactive iodine or external beam radiation therapy in all but 2 of the 13 cases (85%). Mortality in all 13 cases was from a variety of second malignancies (9 separate types) that were not thyroid cancers.

It is clear from these data that in order to improve on long-term outcomes we need to develop an effective mechanism by which to stratify patients into low- and high-risk groups. High-risk patients would presumably continue to benefit from this aggressive approach, while low-risk patients might be effectively treated with less aggressive measures that have a lower risk of life-long complications. To date however, no simple strategy has been proven to predict outcome in individual patients. We attempted to use immunohistochemical and molecular markers to identify individual subjects who would develop recurrent disease. We selected these markers based on their association with an increased risk of recurrence across our entire study population. Despite their statistical significance across groups, several patients with low-risk markers developed recurrent disease while several patients with high-risk markers did not [94, 111–113].

Among adult thyroid cancers, several histological features are associated with a more aggressive clinical course. These have not usually been segregated for independent analysis in studies of childhood thyroid cancer but we believe their importance in adults warrants a discussion in this chapter with the hope that future pediatric studies will identify individual subtypes of PTC and determine the risks for each individual subtype. This stratification might be useful for selecting high- and low-risk subjects.

Thyroid Tumor Pathology

In the last edition of the classification of thyroid tumors from the World Health Organization (2004), thyroid tumors are separated into 3 groups: thyroid carcinomas, thyroid adenomas and other thyroid tumors (table 1) [114–116].

PTC is a malignant epithelial tumor with evidence of follicular cell differentiation and distinctive nuclear features such as clear nuclei and intranuclear inclusions [22]. In children, typical PTC comprise 40% of all cases, the follicular variant of PTC (FVPTC) accounts for 30% of cases, the solid variant is seen in 10–35% of cases and the mixed variant is found in 20–30%. The solid variant

Table 1. Histological classification of thyroid tumors according to the World Health Organization (2004)

Thyroid carcinomas	*Thyroid adenomas and related tumors*
Papillary carcinoma	FA
Follicular carcinoma	Hyalinizing trabecular tumor
Poorly differentiated carcinoma	
Undifferentiated carcinoma	*Other thyroid tumors*
Squamous cell carcinoma	Teratoma
Mucoepidermoid carcinoma	Primary lymphoma and plasmacytoma
Sclerosing mucoepidermoid carcinoma with eosinophilia	Ectopic thymoma
Mucinous carcinoma	Angiosarcoma
Medullary carcinoma	Smooth muscle tumors
Mixed medullary and follicular carcinoma	Peripheral nerve sheath tumors
Spindle cell tumor with thymus-like differentiation	Paraganglioma
Carcinoma showing thymus-like differentiation	Solitary fibrous tumor
	Follicular dendritic cell tumor
	Langherans cell histiocytosis
	Secondary tumors

is thought to occur more commonly in children particularly in radiation-induced tumors. By immunohistochemical staining, the RET oncogene has been detected in 80% of childhood PTC with solid patters of growth. The ret/PTC-3 translocation is most frequent in radiation-induced PTC in children, whereas ret/PTC-1 is more frequently detected in sporadic PTC of children and adults [47, 68].

The mixed variant of PTC in children is most often defined by the presence of equivalent areas of solid and follicular growth containing only occasional loci of papillary structure. The columnar variant is uncommon but is more aggressive in adults and possibly in children. Cribriform PTC and PTC with focal, insular, squamous, or anaplastic components are also very uncommon. The diffuse sclerosing, oxyphilic cell, clear cell, and encapsulated variants of PTC are rare in children. The true combined papillary-medullary variant of PTC is exceedingly rare and its significance is debated.

With few exceptions, correlations between molecular profiles and histology have been inconsistent. The TRK rearrangement occurs in 10% of PTC, while mutations at codon 61 of NRAS are found exclusively in the FVPTC [71]. However, the ret/PTC rearrangements and BRAF mutations are the most frequent molecular events found in PTC with variable prevalence between 10 and 80% among studies [68, 73]. Some of these variations are correlated with the histology and epidemiology of the tumors (ret/PTC-3 and radiation-induced disease for example), while others are probably due to differences in methodology [64–66].

PTC with solid patterns of growth are more likely to develop metastases (75% of cases). However, other features that are associated with low-risk disease in adults such as papillary microcarcinomas (PTC measuring <1 cm) are frequently associated with invasive or metastatic disease in children [87].

Thyroid follicular carcinoma is characterized by the presence of invasion that penetrates through the tumor capsule and/or directly into the blood vessels [23]. In adults, thyroid FTC in nonendemic regions comprise 5–10% of malignant thyroid tumors, but in areas of iodine deficiency the incidence of FTC increases to 25–40%. In children, FTC are generally uncommon and have been reported in as few as 3% to as many as 20% of all cases.

Microscopically, FTC are composed of follicles (microfollicles or occasionally macrofollicles), or trabeculae. Solid areas may also be noted. These lesions tend to be unifocal since they spread by vascular channels and do not invade the lymphatics. They metastasize by a hematogenous route. The oncocytic and clear cell variants remain the two main variants of FTC. The oncocytic variant of FTC accounts for 3–4% of thyroid malignancies. Some studies suggest a primary role for mitochondrial abnormalities in their genesis.

FTC may be subdivided into two distinct pathological types. The first is FTC in which gross invasion is noted at the time of surgery, and the second is a macroscopically encapsulated variant. Another form, the 'grossly encapsulated angioinvasive' form, has been recently introduced along with the classical 'minimally' and 'widely invasive' forms of FTC. This was done to distinguish encapsulated FTC showing minimal capsular invasion from encapsulated FTC showing vascular invasion. Among these encapsulated variants, the risk of local recurrence or distant metastasis is higher when vascular invasion is present.

Cytogenetic and comparative genomic hybridization studies suggest involvement of genes located on chromosomes 2, 3p, 6, 7q, 8, 9, 10q, 11, 13q, 17p and 22 in the genesis of FTC [117,118]. Rearrangements of the peroxisome proliferator-activated receptor gamma (*PPARγ*) are found in 25–50% of FTC [119]. The most common recombinant gene gives rise to the *PAX8-PPARγ* rearrangement which is commonly found in low-stage angioinvasive FTC. Mutations of the *RAS* genes are found in 20–50% of FTC [120]. The most common are activating mutations in codon 61 of *H-* and *N-RAS*.

Most thyroid cancers of follicular origin are categorized as well-differentiated (papillary or follicular) tumors with excellent prognosis. A small number are classified as anaplastic or undifferentiated and have poor prognosis [121]. It seems reasonable that a tumor of intermediate grade should exist. These, poorly differentiated thyroid carcinomas present limited evidence of follicular cell differentiation and are classified as intermediate between differentiated and anaplastic tumors. Several lesions may be candidates for this intermediate group. Poorly differentiated papillary and follicular carcinomas tend to show solid, trabecular, or scirrhous areas usually found in tumors that are Tg positive and therefore of follicular derivation. This class of tumors also includes solid, insular, trabecular FTC and poorly differentiated PTC. They are thought to derive from preexisting differentiated FTC and PTC, but may also develop 'de novo'.

Alterations of *TP53*, *PTEN* and β-catenin genes are implicated in the progression from well-differentiated to poorly differentiated FTC [122–123]. Cytogenetic studies of poorly differentiated carcinomas show complex chromosomal aberrations [125]. Mutations of *RAS* occur in 50% of cases and principally involve N-*RAS* at codon 61 [59, 60, this study]. In contrast, ret/PTC or *TRK* rearrangements are uncommon. Mutations of p53 are found in 50% of the cases and β-catenin in 0–30%. The 5-year survival (50%) is significantly lower for poorly differentiated carcinomas than for differentiated follicular and papillary carcinomas.

Although rare in children, metastatic tumors to the thyroid should always be considered when an unusual histology pattern is encountered. Immunostaining for Tg and/or calcitonin may be useful in distinguishing tumors of thyroidal origin from metastatic disease. Metastatic carcinomas to the thyroid usually represent carcinomas of the kidney, colon, lung, breast, and melanoma.

Anaplastic carcinomas tend to occur in elderly patients (especially women) who have a long history of goiter [121]. The tumors present no evidence of follicular cell differentiation and are categorized as small- or large-cell types. The lesions tend to be rapidly growing and are histologically composed of spindled and sometimes epithelial cells. Many histological variants have been described: epithelial-resembling large-cell lung carcinoma and spindled with features of hemangiopericytoma, malignant fibrous histiocytoma, rhabdomyosarcoma, and osteoclastoma. Extensive pleomorphism and necrosis are common and extrathyroidal spread is usually seen.

Immunohistochemical analysis may help distinguish anaplastic carcinoma from lymphoma. Leukocyte common antigen and/or immunoglobulin stains may establish the lymphoid derivation of a neoplasm. Tumors classified as anaplastic carcinomas often contain keratin or epithelial membrane antigen. However, one third of such lesions are so undifferentiated that no markers can be identified by immunohistochemistry.

MTC are malignant thyroid tumors showing C cell differentiation [126]. This category includes classical forms and numerous variants, i.e. papillary, follicular, spindle cell, oncocytic variant, and others. In most of these cases, the tumors are heterogeneous and careful search on multiple blocks will find some more typical areas, allowing the diagnosis. Immunohistochemical staining of chromogranin, calcitonin and carcinoembryonic antigen is useful in less typical variants. Staining for Tg, Ki 67 (Mib1), and PS100 may also help to distinguish MTC (TCT+, TG−, PS100−) from poorly differentiated follicular carcinomas (TCT−, TG+), hyalinizing trabecular adenoma (TCT−, TG+, Mib1+), or paragangliomas (TCT−, PS100+ in sustentacular cells). Molecular genetic typing in sporadic forms may show a loss of heterozygosity involving chromosomes 1, 3, 11, 13, 17 or 22 [127]. Somatic mutation of RET M918T is found in 20–80% of all cases. Heritable forms include familial MTC, and MTC associated with MEN2A and MEN2B and are discussed in detail in other chapters.

FA are benign encapsulated thyroid tumors consisting of follicular epithelial cells [128]. The FA is the second most common benign nodular lesion in children (second only to nodular goiter) accounting for up to 30% of all cases. FA are more commonly seen in girls than boys (the female:male ratio is 4.5:1). Grossly the adenoma or adenomatous follicular nodule is a solitary lesion demarcated from the surrounding thyroid gland by a capsule of fibrous tissue. Microscopically, these lesions are usually composed of follicles of varying sizes with macrofollicles, microfollicles, or trabeculae being present. However, some nodules are uniform in composition. Thyroid adenomas include macro- and microfollicular, adenolipoma, oxyphilic, clear cell, mucin-producing, and papillary subtypes. The surrounding thyroid may show compression atrophy or may conversely appear normal.

Most adenomas represent monoclonal proliferations, i.e. they are true neoplasms. Isoenzyme histochemical studies in animals have shown monoclonal derivation in solitary nodules and polyclonal proliferations in multinodular goiters. Clonal cytogenetic aberrations are found in 45% of FA. The most common are trisomy 7, translocations 19q13 and 2p21, and deletions on 3p, 10, 13 and 19 [129]. Toxic adenomas frequently contain mutations in the TSH receptor or Gsα while Ras mutations are found in micro-FA [130]. In these tumors, the presence of cellular atypia is correlated with mutation in codon 61 of NRAS, and the presence of incomplete nuclear features of papillary cancers correlates with the presence of ret/PTC rearrangements. PAX8/PPARγ rearrangements are found in some micro-FA. Chromosomal DNA imbalance and mitochondrial DNA polymorphism exist in all types of oncocytic tumors.

The actual nature of trabecular hyalinizing tumors is still debated, since a high rate of ret/PTC rearrangements have been described in these tumors and the nuclei possess some characteristics of papillary cancers. However, malignant cases are so infrequent that their reality remains questionable.

According to the Post-Chernobyl Thyroid Pathologists group, use of the highly controversial term ‘atypical adenoma’ is discouraged and should be replaced by ‘tumor of uncertain malignant potential’ when a tumor exhibits incomplete features of PTC or disputable capsular invasion [131]. The distinction of these tumors from FVPTC has triggered much debate. In part this is because recognition of the so-called 'PAP nuclei' and the distinction of FVPTC from tumors possessing only limited nuclear features of PTC is poorly reproducible and subject to great interobserver variation (nearly 60%) [132]. Immunohistochemical detection of malignant markers can improve the cytological diagnosis of thyroid tumors on FNA, but has little utility for the final histological diagnosis of malignancy. Indeed, HBME1 staining and TPO alterations are not fully specific for malignancy, whereas galectin is often absent from malignant follicular tumors [133, 134]. It is recommended by most authors to make a diagnosis of FVPTC when more than 50% of nuclei are typically papillary or when some foci are entirely made of cells possessing typical papillary nuclei.

Follow-Up

The optimal means and frequency with which to detect persistent or recurrent disease in children are not well defined. Hung and Sarlis [31] suggest that radioactive iodine whole body scan (RAI-WBS) and serum Tg levels should be determined at 6-month intervals for the first 18 months after surgery. Persistent or recurrent disease should be treated at 6-month intervals. Once the patient is free from disease, thyroid US and chest computerized tomography (CT) should also be performed. If all studies are negative, serum Tg and RAI-WBS could be performed at 3 and 5 years and if the patient is free from disease at 11 years and 6 months, annual unstimulated serum Tg could be used along with RAI-WBS and stimulated Tg at intervals of every 5 years.

This approach is commonly used but might not take full advantage of thyroid US. Furthermore, the absolute value of serum Tg levels that indicate disease of sufficient quantity to benefit from treatment is not well defined. Antonelli et al. [135] compared the results of RAI-WBS and thyroid US in 35 children and the results of thyroid US and serum Tg in 29 children. Three out of 35 (8.6%) had negative RAI-WBS but positive US and 6/35 (17.1%) had positive RAI-WBS but negative US. Likewise, 5/29 (17.2%) had negative Tg but positive US while 10/29 (34.4%) had positive Tg but negative US. Overall, 8 (23%) patients had disease not detected by Tg or RAI-WBS and 16 (46%) patients had disease not detected by US. Their observations suggest that most patients will have disease that can be detected by RAI-WBS, serum Tg, or US, but that a few will have disease detected by only one of these tests. For that reason, patients might benefit from all three procedures at regular intervals.

The value of chest CT in the detection of residual or recurrent disease is not strongly supported. Bal et al. [71] evaluated 28 patients who were <20 years of age (range: 6–20 years, mean: 13.9 years) using whole body scintigraphy (2–3 mCi), chest radiograph and chest CT. Almost all (93%) had undergone near-total thyroidectomy and cervical node dissection. The patients were then treated with radioactive iodine and posttherapy images were obtained. All 28 patients had disease visualized on the posttherapy images but only 25–30% of chest radiographs and chest CT identified disease and the standard 2–3 mCi whole body scan found disease in only slightly more than half of the patients. These data question the sensitivity of radiographs, chest CT and even routine RAI-WBS for the detection of pulmonary metastases in children.

Thyroid hormone withdrawal or recombinant human thyrotropin (rhTSH) stimulation with subsequent determination of the serum Tg level has been increasingly important for the detection of residual or recurrent disease in adults. As of this time, however, rhTSH is not yet approved for use in children and the data pertaining to its utility in this age group are sparse.

However, Pacini et al. [82, 136] compared serum Tg determinations and RAI-WBS for detecting residual disease in adults following thyroid hormone withdrawal. Among the 315 patients with undetectable serum Tg levels, 225 (71.4%) had negative RAI-WBS and 90 (28.6%) had uptake in the thyroid bed. No local or distant metastases were discovered. Over time, 281 patients (89.2%) showed complete remission and only 2 patients (0.6%) developed local recurrence. Their data suggest that undetectable serum Tg levels stimulated by thyroid hormone withdrawal are predictive of complete remission [82].

Cailleux et al. [137] found that serum Tg levels above 10 ng/ml (off thyroid hormone suppression) are highly predictive of residual disease. Serum Tg levels >10 ng/ml were found in 15/256 patients (6%). Three of these patients had disease outside the thyroid bed, 1 (11%) had pulmonary metastases, and 1 (11%) had cervical lymph node involvement. Overall, 5 of these 15 patients (33%) had disease outside the thyroid bed [137].

Mazzaferri and Kloos [138] showed that rhTSH-stimulated serum Tg was a sensitive method by which to detect disease. Twenty out of 107 (19%) patients had rhTSH-stimulated Tg >2 ng/ml. Eleven of them (10%) had disease in the lung or lymph nodes. These findings have been confirmed by multiple studies. Haugen et al. [139] found that rhTSH-stimulated Tg levels <4 ng/ml rarely required additional evaluation while Robbins et al. [140] found that a rhTSH-stimulated Tg ≤2 ng/ml had a 91.7% negative predictive value.

What remains somewhat ambiguous from these data is the significance of a rhTSH-stimulated serum Tg value between 2 and 10 ng/ml in the setting of a negative diagnostic whole body scan. Wang et al. [141] evaluated the response to therapy of patients with rhTSH-stimulated serum Tg between 2 and 10 ng/ml. Over time, the majority of patients had stable or decreasing rhTSH-stimulated Tg values when followed without additional radioactive iodine or surgical intervention. Empiric therapy with 150–200 mCi of radioactive iodine did not appear to result in a more substantial decline in these low level serum Tg levels. To date, however, no study has defined the absolute lower limit of serum Tg that would indicate disease of sufficient magnitude to warrant therapy in a child. It is possible that these data pertain to children, but the possibility of illegitimate Tg transcription and false-positive serum Tg values, particularly at the lower limits of detection, remain a concern. This has led to the concept of treating to a negative whole body scan and continual follow-up of serum Tg values [142].

Despite this controversy, we believe that serial Tg measurements are a sensitive method by which to assess disease status [143]. Patients who have been treated with total thyroidectomy and radioactive iodine ablation should have undetectable serum Tg ($\leq$0.5 ng/ml) while on thyroid hormone suppression [82]. In the absence of circulating Tg antibodies, recurrent disease should be suspected in any child with detectable serum Tg [40]. Whether or not to treat individual children based on serum Tg alone will remain an individual decision based on risk of recurrence, magnitude of serum Tg, and the history of prior response to therapy. We presume that the patients with stable but persistent disease reported by La Quaglia et al. [144] would have persistent elevations in serum Tg but whether or not these remain stable over time remains unknown.

For children and adolescents, life-long follow-up is required. Use of rhTSH to stimulate Tg production in lieu of thyroid hormone withdrawal could be an ideal means by which to follow these patients. However, insufficient data currently exist to allow us to determine if rhTSH-stimulated serum Tg levels are as effective for detection of disease as are our current protocols. These are generally based on thyroid hormone withdrawal for the stimulation of serum Tg and the preparation for a RAI-WBS. Thyroid US is also being increasingly used for follow-up. Iorcansky et al. [145] have recently published initial data regarding the safety of rhTSH stimulation in children. They examined the serum TSH levels obtained following two injections of rhTSH (0.9 mg each injection which is the adult dose) given 24 h apart to 19 children and adolescents. Their data show that at the time of radioactive iodine administration, serum TSH levels (134 $\pm$ 75 mIU/l) were similar to those obtained in children following thyroid hormone withdrawal (188 $\pm$ 118 mIU/l, $p = 0.07$) and similar to those in adults following rhTSH stimulation (105 $\pm$ 43 and 124 $\pm$ 59 mIU/l from two different centers). There were no untoward effects from the adult dose of rhTSH in these children.

The optimal frequency with which to perform stimulated serum Tg testing (either in response to thyroid hormone withdrawal or rhTSH) is unknown for children but it is our clinical practice to assess serum Tg responses on an annual basis and to treat patients with persistent or recurrent disease based on the results of these annual tests. We stratify our patients according to the risk of detecting disease based on age (<10 years are high risk), tumor size (>2 cm are high risk), previous disease in the cervical nodes (high risk), or pulmonary metastases (high risk). Patients at high risk are prepared by conventional thyroid hormone withdrawal in anticipation of a requirement to treat with radioactive iodine. The latter will be expedited by preparing the patients with conventional thyroid hormone withdrawal. Low-risk patients (when old enough) could be prepared with rhTSH stimulation in anticipation of finding no active disease.

The requirement for annual Tg stimulation testing, RAI-WBS and US will be relaxed over time but the optimal time at which to lengthen the intervals between examinations is not yet clear. Welch Dinauer et al. [52] found that 90% of all her patients who developed recurrent disease did so within 7 years after diagnosis. Other studies have shown equal probability of recurrence in the first and second decades after diagnosis [146]. Again, one might revert to risk stratification to help identify those patients at higher risk of recurrence. These might benefit from longer annual surveillance.

One question commonly raised by parents is whether or not PTC are hereditary and whether not other children in the family are at risk. Overall about 5% of PTC are inherited through a dominant mode of transmission [147]. PTC can also be the presenting malignancy in familiar adenomatoid polyposis, Gardner's syndrome (an association of intestinal tumors, desmoid tumors, lipomas, and epidermoid cysts) or Cowden's syndrome (an association of multiple hamartomas, breast cancer, colon cancer, and nodular goiter) [40]. Another interesting and provocative report by Memon et al. [148] evaluated 313 cases and case controls. Seventy-eight (24%) of the cases had a positive family history of benign thyroid disease compared to only 12.8% of case controls. Overall, there was a 2-fold increased risk of thyroid cancer if the mother or sister had benign thyroid disease, but there was no correlation between benign thyroid disease and other cancers such as breast cancer. Overall, this suggests a possible link or common association between benign and malignant thyroid diseases.

Summary

The American Thyroid Association (ATA) Taskforce recently published guidelines for treatment of patients with differentiated thyroid cancer [149]. We believe children are sufficiently unique to warrant consideration of these treatment protocols on an individual basis. Thyroid cancers are uncommon pediatric tumors but an updated understanding of the molecular biology and treatment of these conditions will assist the clinician in selecting specific treatment and follow-up programs for individual patients. In general, children with thyroid cancers may have a modest decline in life expectancy, but they continue to be at risk of recurrent disease through many decades. Optimal surveillance techniques and intervals have yet to be determined and it is our hope that this chapter will stimulate interest in collaborative studies designed to answer questions such as the optimal level of TSH suppression that might be appropriate for children with PTC or FTC, the optimal interval for serum Tg measurements and whole body scans, and the use of rhTSH stimulation.

The opinions or assertions contained herein are the private views of the authors and are not to be construed as official or to reflect the opinions of the Uniformed Services University of the Health Sciences, Walter Reed Army Medical Center, the Department of the Army, or the Department of Defense.

References

1 Winship T, Rosvall RV: Childhood thyroid carcinoma. Cancer 1961;14:734–743.
2 Wartofsky L: The thyroid nodule; in Wartofsky L (ed): Thyroid Cancer: a Comprehensive Guide to Clinical Management. Totowa, Humana Press, 2000, pp 3–7.
3 Moretti F, Nanni S, Pontecorvi A: Molecular pathogenesis of thyroid nodules and cancer. Baillieres Best Pract Res Clin Endocrinol Metab 2000;14:517–539.
4 Hopwood NJ, Kelch RP: Thyroid masses: approach to diagnosis and management in childhood and adolescence. Pediatr Rev 1993;14:481–487.
5 Hung W: Nodular thyroid disease and thyroid carcinoma. Pediatr Ann 1992;21:50–57.
6 Vierhapper H, Niederle B, Bieglmayer C, Kaserer K, Baumgartner-Parzer S: Early diagnosis and curative therapy of medullary thyroid carcinoma by routine measurement of serum calcitonin in patients with thyroid disorders. Thyroid 2005;15:1267–1272.
7 Lugo-Vicente H, Ortiz VN: Pediatric thyroid nodules: insights in management. Bol Asoc Med P R 1998;90:74–78.
8 Mircescu H, Parma J, Huot C, Deal C, Oligny LL, Vassart G, Van Vliet G: Hyperfunctioning malignant thyroid nodule in an 11-year-old girl: pathologic and molecular studies. J Pediatr 2000;137:585–587.
9 Corrias A, Einaudi S, Chiorboli E, Weber G, Crino A, Andreo M, Cesaretti G, de Sanctis L, Messina MF, Segni M, Cicchetti M, Vigone M, Pasquino AM, Spera S, de Luca F, Mussa GC, Bona G: Accuracy of fine needle aspiration biopsy of thyroid nodules in detecting malignancy in childhood: comparison with conventional clinical, laboratory, and imaging approaches. J Clin Endocrinol Metab 2001;86:4644–4648.
10 James EM, Charboneau JW, Hay ID: The thyroid; in Rumack CM, Wilson SR, Charboneau JW (eds): Diagnostic Ultrasound. St Louis, Mosby Year Book, 1991, vol 1, pp 507–523.
11 De Nicola H, Szejnfeld J, Logullo AF, Wolosker AM, Souza LR, Chiferi V Jr: Flow pattern and vascular resistive index as predictors of malignancy risk in thyroid follicular neoplasms. J Ultrasound Med 2005;24:897–904.
12 Tan GH, Gharib H, Reading CC: Solitary thyroid nodule. Comparison between palpation and ultrasonography. Arch Intern Med 1995;155:2418–2423.
13 Mazzaferri EL: Thyroid cancer in thyroid nodules: finding a needle in the haystack. Am J Med 1992;93:359–362.
14 Mazzaferri EL: Management of a solitary thyroid nodule. N Engl J Med 1993;328:553–559.
15 Gharib H: Fine-needle aspiration biopsy of thyroid nodules: advantages, limitations, and effect. Mayo Clin Proc 1994;69:44–49.
16 Gharib H, Goellner JR: Fine-needle aspiration biopsy of the thyroid: an appraisal. Ann Intern Med 1993;118:282–289.

17 Burch HB, Burman KD, Reed HL, Buckner L, Raber T, Ownbey JL: Fine needle aspiration of thyroid nodules. Determinants of insufficiency rate and malignancy yield at thyroidectomy. Acta Cytol 1996;40:1176–1183.
18 Oertel YC: Fine-needle aspiration and the diagnosis of thyroid cancer. Endocrinol Metab Clin North Am 1996;25:69–91.
19 Oertel YC: The thyroid nodule: fine needle aspiration biopsy; in Wartofsky L (ed): Thyroid Cancer: a Comprehensive Guide to Clinical Management. Totowa, Humana Press, 2000, pp 35–38.
20 Degnan BM, McClellan DR, Francis GL: An analysis of fine-needle aspiration biopsy of the thyroid in children and adolescents. J Pediatr Surg 1996;31:903–907.
21 Raab SS, Silverman JF, Elsheikh TM, Thomas PA, Wakely PE: Pediatric thyroid nodules: disease demographics and clinical management as determined by fine needle aspiration biopsy. Pediatrics 1995;95:46–49.
22 Oertel J, Oertel Y: Papillary carcinoma; in Wartofsky L (ed): Thyroid Cancer: a Comprehensive Guide to Clinical Management. Totowa, Humana Press, 2000, pp 193–208.
23 Oertel J, Oertel Y: Pathology of follicular thyroid cancer; in Wartofsky L (ed): Thyroid Cancer: a Comprehensive Guide to Clinical Management. Totowa, Humana Press, 2000, pp 289–296.
24 Orija IB, Hamrahian AH, Reddy SS: Management of nondiagnostic thyroid fine-needle aspiration biopsy: survey of endocrinologists. Endocr Pract 2004;10:317–323.
25 Haugen BR, Nawaz S, Markham N, Hashizumi T, Shroyer AL, Werness B, Shroyer KR: Telomerase activity in benign and malignant thyroid tumors. Thyroid 1997;7:337–342.
26 Lerma E, Mora J: Telomerase activity in 'suspicious' thyroid cytology. Cancer 2005;105:492–497.
27 Kammori M, Takubo K, Nakamura K, Furugouri E, Endo H, Kanauchi H, Mimura Y, Kaminishi M: Telomerase activity and telomere length in benign and malignant human thyroid tissues. Cancer Lett 2000;159:175–181.
28 Straight A, Patel A, Fenton C, Dinauer C, Tuttle RM, Francis G: Thyroid carcinomas that express telomerase follow a more aggressive clinical course for children and adolescents. J Endocrinol Invest 2002;25:302–308.
29 Papotti M, Volante M, Saggiorato E, Deandreis D, Veltri A, Orlandi F: Role of galectin-3 immunodetection in the cytological diagnosis of thyroid cystic papillary carcinoma. Eur J Endocrinol 2002;147:515–521.
30 Orlandi A, Puscar A, Capriata E, Fideleff H: Repeated fine-needle aspiration of the thyroid in benign nodular thyroid disease: critical evaluation of long-term follow-up. Thyroid 2005;15:274–278.
31 Hung W, Sarlis NJ: Current controversies in the management of pediatric patients with well-differentiated nonmedullary thyroid cancer: a review. Thyroid 2002;12:683–702.
32 Gharib H: Changing concepts in the diagnosis and management of thyroid nodules. Endocrinol Metab Clin North Am 1997;26:777–800.
33 Gharib H, Mazzaferri EL: Thyroxine suppressive therapy in patients with nodular thyroid disease. Ann Intern Med 1998;128:386–394.
34 Giuffrida D, Gharib H: Controversies in the management of cold, hot, and occult thyroid nodules. Am J Med 1995;99:642–650.
35 Sdano MT, Falciglia M, Welge JA, Steward DL: Efficacy of thyroid hormone suppression for benign thyroid nodules: meta-analysis of randomized trials. Otolaryngol Head Neck Surg 2005;133: 391–396.
36 Papini E, Petrucci L, Guglielmi R, Panunzi C, Rinaldi R, Bacci V, Crescenzi A, Nardi F, Fabbrini R, Pacella CM: Long-term changes in nodular goiter: a 5-year prospective randomized trial of levothyroxine suppressive therapy for benign cold thyroid nodules. J Clin Endocrinol Metab 1998;83:780–783.
37 Cooper DS: Clinical review 66: thyroxine suppression therapy for benign nodular disease. J Clin Endocrinol Metab 1995;80:331–334.
38 Faber J, Galloe AM: Changes in bone mass during prolonged subclinical hyperthyroidism due to L-thyroxine treatment: a meta-analysis. Eur J Endocrinol 1994;130:350–356.
39 Nuzzo V, Lupoli G, Esposito Del Puente A, Rampone E, Carpinelli A, Del Puente AE, Oriente P: Bone mineral density in premenopausal women receiving levothyroxine suppressive therapy. Gynecol Endocrinol 1998;12:333–337.

40 Bauer AJ, Tuttle RM, Francis G: Thyroid nodules and thyroid carcinoma in children; in Pescovitz O, Eugster E (eds): Pediatric Endocrinology: Mechanisms, Manifestations, and Management. Philadelphia, Lippincott, 2004.
41 Ehrhardt O: Zur Anatomie und Klinik der Struma maligna. Beitr Klin Chir 1902;35:343–364.
42 DeGroot L, Paloyan E: Thyroid carcinoma and radiation. A Chicago endemic. JAMA 1973;225: 487–491.
43 Mangano JJ: A post-Chernobyl rise in thyroid cancer in Connecticut, USA. Eur J Cancer Prev 1996;5:75–81.
44 Ries LAG, Smith M, Gurney JG, Linet M, Tamra T, Young JL, Bunin GR: Cancer Incidence and Survival among Children and Adolescents: United States SEER Program, 1975–1995. Bethesda, National Cancer Institute, 1999, Publ No 99-4649.
45 Nikiforov Y: Spatial positioning of RET and H4 following radiation exposure leads to tumor development. Scientific World J 2001;1:186–187.
46 Buckwalter TL, Venkateswaran A, Lavender M, La Perle KM, Cho JY, Robinson ML, Jhiang SM: The roles of phosphotyrosines-294, -404, and -451 in RET/PTC1-induced thyroid tumor formation. Oncogene 2002;21:8166–8172.
47 Nikiforov YE, Koshoffer A, Nikiforova M, Stringer J, Fagin JA: Chromosomal breakpoint positions suggest a direct role for radiation in inducing illegitimate recombination between the ELE1 and RET genes in radiation-induced thyroid carcinomas. Oncogene 1999;18:6330–6334.
48 Ron E, Lubin JH, Shore RE, Mabuchi K, Modan B, Pottern LM, Schneider AB, Tucker MA, Boice JD Jr: Thyroid cancer after exposure to external radiation: a pooled analysis of seven studies. Radiat Res 1995;141:259–277.
49 Shore RE: Issues and epidemiological evidence regarding radiation-induced thyroid cancer. Radiat Res 1992;131:98–111.
50 Becker DV, Robbins J, Beebe GW, Bouville AC, Wachholz BW: Childhood thyroid cancer following the Chernobyl accident: a status report. Endocrinol Metab Clin North Am 1996;25:197–211.
51 Nauman J, Wolff J: Iodide prophylaxis in Poland after the Chernobyl reactor accident: benefits and risks. Am J Med 1993;94:524–532.
52 Welch Dinauer CA, Tuttle RM, Robie DK, McClellan DR, Svec RL, Adair C, Francis GL: Clinical features associated with metastasis and recurrence of differentiated thyroid cancer in children, adolescents and young adults. Clin Endocrinol (Oxf) 1998;49:619–628.
53 Hung W, Anderson KD, Chandra RS, Kapur SP, Patterson K, Randolph JG, August GP: Solitary thyroid nodules in 71 children and adolescents. J Pediatr Surg 1992;27:1407–1409.
54 Tubiana M, Schlumberger M, Rougier P, Laplanche A, Benhamou E, Gardet P, Caillou B, Travagli JP, Parmentier C: Long-term results and prognostic factors in patients with differentiated thyroid carcinoma. Cancer 1985;55:794–804.
55 Powers PA, Dinauer CA, Tuttle RM, Francis GL: Treatment of recurrent papillary thyroid carcinoma in children and adolescents. J Pediatr Endocrinol Metab 2003;16:1033–1040.
56 Powers PA, Dinauer CA, Tuttle RM, Robie DK, McClellan DR, Francis GL: Tumor size and extent of disease at diagnosis predict the response to initial therapy for papillary thyroid carcinoma in children and adolescents. J Pediatr Endocrinol Metab 2003;16:693–703.
57 Brink JS, van Heerden JA, McIver B, Salomao DR, Farley DR, Grant CS, Thompson GB, Zimmerman D, Hay ID: Papillary thyroid cancer with pulmonary metastases in children: long-term prognosis. Surgery 2000;128:881–887.
58 LaQuaglia M, Black T, Holcomb G, Sklar C, Azizkhan R, Haase G, Newman K: Differentiated thyroid cancer: clinical characteristics, treatment, and outcome in patients under 21 years of age who present with distant metastases. A report from the Surgical Discipline Committee of the Children's Cancer Group. J Pediatr Surg 2000;35:955–959.
59 Bertherat J, Groussin L, Sandrini F, Matyakhina L, Bei T, Stergiopoulos S, Papageorgiou T, Bourdeau I, Kirschner LS, Vincent-Dejean C, Perlemoine K, Gicquel C, Bertagna X, Stratakis CA: Molecular and functional analysis of PRKAR1A and its locus (17q22-24) in sporadic adrenocortical tumors: 17q losses, somatic mutations, and protein kinase A expression and activity. Cancer Res 2003;63:5308–5319.
60 Nikiforova MN, Lynch RA, Biddinger PW, Alexander EK, Dorn GW 2nd, Tallini G, Kroll TG, Nikiforov YE: RAS point mutations and PAX8-PPAR gamma rearrangement in thyroid tumors:

evidence for distinct molecular pathways in thyroid follicular carcinoma. J Clin Endocrinol Metab 2003;88:2318–2326.
61 Hara H, Fulton N, Yashiro T, Ito K, DeGroot LJ, Kaplan EL: N-ras mutation: an independent prognostic factor for aggressiveness of papillary thyroid carcinoma. Surgery 1994;116: 1010–1016.
62 Nikiforov YE, Nikiforova MN, Gnepp DR, Fagin JA: Prevalence of mutations of ras and p53 in benign and malignant thyroid tumors from children exposed to radiation after the Chernobyl nuclear accident. Oncogene 1996;13:687–693.
63 Fenton C, Anderson J, Lukes Y, Dinauer CA, Tuttle RM, Francis GL: Ras mutations are uncommon in sporadic thyroid cancer in children and young adults. J Endocrinol Invest 1999;22:781–789.
64 Pacini F, Agate L, Elisei R, Capezzone M, Ceccarelli C, Lippi F, Molinaro E, Pinchera A: Outcome of differentiated thyroid cancer with detectable serum Tg and negative diagnostic (131)I whole body scan: comparison of patients treated with high (131)I activities versus untreated patients. J Clin Endocrinol Metab 2001;86:4092–4097.
65 Nikiforov Y, Gnepp DR, Fagin JA: Thyroid lesions in children and adolescents after the Chernobyl disaster: implications for the study of radiation tumorigenesis. J Clin Endocrinol Metab 1996;81: 9–14.
66 Thomas GA, Bunnell H, Cook HA, Williams ED, Nerovnya A, Cherstvoy ED, Tronko ND, Bogdanova TI, Chiappetta G, Viglietto G, Pentimalli F, Salvatore G, Fusco A, Santoro M, Vecchio G: High prevalence of RET/PTC rearrangements in Ukrainian and Belarussian post-Chernobyl thyroid papillary carcinomas: a strong correlation between RET/PTC3 and the solid-follicular variant. J Clin Endocrinol Metab 1999;84:4232–4238.
67 Learoyd DL, Messina M, Zedenius J, Guinea AI, Delbridge LW, Robinson BG: RET/PTC and RET tyrosine kinase expression in adult papillary thyroid carcinomas. J Clin Endocrinol Metab 1998;83:3631–3635.
68 Fenton CL, Lukes Y, Nicholson D, Dinauer CA, Francis GL, Tuttle RM: The ret/PTC mutations are common in sporadic papillary thyroid carcinoma of children and young adults. J Clin Endocrinol Metab 2000;85:1170–1175.
69 Kimura ET, Nikiforova MN, Zhu Z, Knauf JA, Nikiforov YE, Fagin JA: High prevalence of BRAF mutations in thyroid cancer. Cancer Res 2003;63:1454–1457.
70 Soares P, Trovisco V, Rocha AS, Lima J, Castro P, Preto A, Maximo V, Botelho T, Seruca R, Sobrinho-Simoes M: BRAF mutations and RET/PTC rearrangements are alternative events in the etiopathogenesis of PTC. Oncogene 2003;22:4578–4580.
71 Bal CS, Padhy AK, Kumar A: Clinical features of differentiated thyroid carcinoma in children and adolescents from a sub-Himalayan iodine-deficient endemic zone. Nucl Med Commun 2001;22: 881–887.
72 Xing M, Vasko V, Tallini G, Larin A, Wu G, Udelsman R, Ringel MD, Ladenson PW, Sidransky D: BRAF T1796A transversion mutation in various thyroid neoplasms. J Clin Endocrinol Metab 2004;89:1365–1368.
73 Penko K, Livezey J, Fenton C, Patel A, Nicholson D, Flora M, Oakley K, Tuttle RM, Francis G: BRAF mutations are uncommon in papillary thyroid cancer of young patients. Thyroid 2005;15: 320–325.
74 Fenton CL, Patel A, Burch HB, Tuttle RM, Francis GL: Nuclear localization of thyroid transcription factor-1 correlates with serum thyrotropin activity and may be increased in differentiated thyroid carcinomas with aggressive clinical course. Ann Clin Lab Sci 2001;31:245–252.
75 Civitareale D, Lonigro R, Sinclair AJ, Di Lauro R: A thyroid-specific nuclear protein essential for tissue-specific expression of the thyroglobulin promoter. EMBO J 1989;8:2537–2542.
76 Heldin NE, Westermark B: The molecular biology of the human anaplastic thyroid carcinoma cell. Thyroidology 1991;3:127–131.
77 Scouten WT, Patel A, Terrell R, Burch HB, Bernet VJ, Tuttle RM, Francis GL: Cytoplasmic localization of the paired box gene, Pax-8, is found in pediatric thyroid cancer and may be associated with a greater risk of recurrence. Thyroid 2004;14:1037–1046.
78 Patel A, Jhiang S, Dogra S, Terrell R, Powers PA, Fenton C, Dinauer CA, Tuttle RM, Francis GL: Differentiated thyroid carcinoma that express sodium-iodide symporter have a lower risk of recurrence for children and adolescents. Pediatr Res 2002;52:737–744.

79 Gupta S, Patel A, Folstad A, Fenton C, Dinauer CA, Tuttle RM, Conran R, Francis GL: Infiltration of differentiated thyroid carcinoma by proliferating lymphocytes is associated with improved disease-free survival for children and young adults. J Clin Endocrinol Metab 2001;86:1346–1354.
80 Modi J, Patel A, Terrell R, Gupta S, Fenton C, Tuttle R, Francis G: Papillary thyroid carcinoma from children and adolescents contain a mixture of lymphocytes. J Clin Endocrinol Metab 2003;88: 4418–4425.
81 Sugg SL, Ezzat S, Rosen IB, Freeman JL, Asa SL: Distinct multiple RET/PTC gene rearrangements in multifocal papillary thyroid neoplasia. J Clin Endocrinol Metab 1998;83:4116–4122.
82 Pacini F, Capezzone M, Elisei R, Ceccarelli C, Taddei D, Pinchera A: Diagnostic 131-iodine whole-body scan may be avoided in thyroid cancer patients who have undetectable stimulated serum Tg levels after initial treatment. J Clin Endocrinol Metab 2002;87:1499–1501.
83 Borson-Chazot F, Causeret S, Lifante JC, Augros M, Berger N, Peix JL: Predictive factors for recurrence from a series of 74 children and adolescents with differentiated thyroid cancer. World J Surg 2004;28:1088–1092.
84 Brink JS, van Heerden JA, McIver B, Salomao DR, Farley DR, Grant CS, Thompson GB, Zimmermann D, Hay ID: Papillary thyroid cancer with pulmonary metastases in children: long-term prognosis. Surgery 2000;128:881–887.
85 Welch Dinauer CA, Tuttle RM, Robie DK, McClellan DR, Francis GL: Extensive surgery improves recurrence-free survival for children and young patients with class I papillary thyroid carcinoma. J Pediatr Surg 1999;34:1799–1804.
86 Demidchik IE, Kontratovich VA: Repeat surgery for recurrent thyroid cancer in children. Vopr Onkol 2003;49:366–369.
87 Chow S, Law S, Mendenhall W, Au S, Yau S, Mang O, Lau W: Differentiated thyroid carcinoma in childhood and adolescence – clinical course and role of radioiodine. Pediatr Blood Cancer 2004;42:176–183.
88 Leboulleux S, Baudin E, Hartl DW, Travagli JP, Schlumberger M: Follicular cell-derived thyroid cancer in children. Horm Res 2005;63:145–151.
89 Reynolds JC: Comparison of I-131 absorbed radiation doses in children and adults: a tool for estimating therapeutic I-131 doses in children; in Robbins JM (ed): Treatment of Thyroid Cancer in Childhood. Bethesda, NIDDK, National Institutes of Health, 1992, pp 127–135.
90 Benua R, Leeper R: A method and rationale for treating thyroid carcinoma with the largest safe dose of I-131; in Meideros-Neto G, Gaitan E (eds): Frontiers in Thyroidology. New York, Plenum, 1986, vol 2, pp 1317–1321.
91 Dorn R, Kopp J, Vogt H, Heidenreich P, Carroll R, Gulec S: Dosimetry-guided radioactive iodine treatment in patients with metastatic differentiated thyroid cancer: largest safe dose using a risk-adapted approach. J Nucl Med 2003;44:451–456.
92 Fenton C, Patel A, Dinauer C, Robie DK, Tuttle RM, Francis GL: The expression of vascular endothelial growth factor and the type 1 vascular endothelial growth factor receptor correlate with the size of papillary thyroid carcinoma in children and young adults. Thyroid 2000;10:349–357.
93 Lennard CM, Patel A, Wilson J, Reinhardt B, Tuman C, Fenton C, Blair E, Francis GL, Tuttle RM: Intensity of vascular endothelial growth factor expression is associated with increased risk of recurrence and decreased disease-free survival in papillary thyroid cancer. Surgery 2001;129:552–558.
94 Patel A, Straight AM, Mann H, Duffy E, Fenton C, Dinauer C, Tuttle RM, Francis GL: Matrix metalloproteinase (MMP) expression by differentiated thyroid carcinoma of children and adolescents. J Endocrinol Invest 2002;25:403–408.
95 DeGroot LJ, Kaplan EL, McCormick M, Straus FH: Natural history, treatment, and course of papillary thyroid carcinoma. J Clin Endocrinol Metab 1990;71:414–424.
96 Causeret S, Lifante JC, Borson-Chazot F, Varcus F, Berger N, Peix JL: Differentiated thyroid carcinoma in children and adolescents: therapeutic strategy according to clinic presentation. Ann Chir 2004;129:359–364.
97 LaQuaglia M, Telander R: Differentiated and medullary thyroid cancer in children and adolescence. Semin Pediatr Surg 1997;6:42–49.
98 Robie DK, Dinauer CW, Tuttle RM, Ward DT, Parry R, McClellan D, Svec R, Adair C, Francis G: The impact of initial surgical management on outcome in young patients with differentiated thyroid cancer. J Pediatr Surg 1998;33:1134–1140.

99 Kowalski LP, Goncalves Filho J, Pinto CA, Carvalho AL, de Camargo B: Long-term survival rates in young patients with thyroid carcinoma. Arch Otolaryngol Head Neck Surg 2003;129:746–749.
100 van Santen HM, Aronson DC, Vulsma T, Tummers RF, Geenen MM, de Vijlder JJ, van den Bos C: Frequent adverse events after treatment for childhood-onset differentiated thyroid carcinoma: a single institute experience. Eur J Cancer 2004;40:1743–1751.
101 Sweeney D, Johnston G: Radioiodine treatment of thyroid cancer; in Wartofsky L (ed): Thyroid Cancer: a Comprehensive Guide to Clinical Management. Totowa, Humana Press, 2000, pp 155–162.
102 Handelsman DJ, Turtle JR: Testicular damage after radioactive iodine (I-131) therapy for thyroid cancer. Clin Endocrinol (Oxf) 1983;18:465–472.
103 Ahmed SR, Shalet SM: Radioactive iodine and testicular damage. N Engl J Med 1984;311:1576.
104 Pacini F, Gasperi M, Fugazzola L, Ceccarelli C, Lippi F, Centoni R, Martino E, Pinchera A: Testicular function in patients with differentiated thyroid carcinoma treated with radioiodine. J Nucl Med 1994;35:1418–1422.
105 Casara D, Rubello D, Saladini G, Piotto A, Pelizzo MR, Girelli ME, Busnardo B: Pregnancy after high therapeutic doses of iodine-131 in differentiated thyroid cancer: potential risks and recommendations. Eur J Nucl Med 1993;20:192–194.
106 Balan KK, Critchley M: Outcome of pregnancy following treatment of well-differentiated thyroid cancer with 131iodine. Br J Obstet Gynaecol 1992;99:1021–1022.
107 Kulakov VI, Sokur TN, Volobuev AI, Tzibulskaya IS, Malisheva VA, Zikin BI, Ezova LC, Belyaeva LA, Bonartzev PD, Speranskaya NV, et al: Female reproductive function in areas affected by radiation after the Chernobyl power station accident. Environ Health Perspect 1993;101 (suppl 2):117–123.
108 Lin JD, Wang HS, Weng HF, Kao PF: Outcome of pregnancy after radioactive iodine treatment for well differentiated thyroid carcinomas. J Endocrinol Invest 1998;21:662–667.
109 Schlumberger M, De Vathaire F, Ceccarelli C, Francese C, Pinchera A, Parmentier C: Outcome of pregnancy in women with thyroid carcinoma. J Endocrinol Invest 1995;18:150–151.
110 Smith MB, Xue H, Takahashi H, Cangir A, Andrassy RJ: Iodine 131 thyroid ablation in female children and adolescents: long-term risk of infertility and birth defects. Ann Surg Oncol 1994;1: 128–131.
111 Patel A, Fenton C, Terrell R, Powers PA, Dinauer C, Tuttle RM, Francis GL: Nitrotyrosine, inducible nitric oxide synthase (iNOS), and endothelial nitric oxide synthase (eNOS) are increased in thyroid tumors from children and adolescents. J Endocrinol Invest 2002;25:675–683.
112 Gydee H, O'Neill JT, Patel A, Bauer AJ, Tuttle RM, Francis G: Differentiated thyroid carcinomas from children and adolescents express insulin-like growth factor-1 (IGF-1) and the IGF-1 receptor (IGF-1-R). Cancers with the most intense IGF-1-R expression may be more aggressive. Pediatr Res 2004;55:1–7.
113 Powers PA, Dinauer CA, Tuttle RM, Francis G: The MACIS score predicts the clinical course of papillary thyroid carcinoma in children and adolescents. J Pediatr Endocrinol Metab 2004;17:339–343.
114 DeLellis R, Lloyd RV, Heitz PU, Eng C: World Health Organization Classification of Tumours. Pathology and Genetics of Tumours of Endocrine Organs. Lyon, IARC Press, 2004.
115 Hedinger C, Williams ED, Sobin LH: The WHO histological classification of thyroid tumors: a commentary on the second edition. Cancer 1989;63:908–911.
116 Wittekind C, Greene F, Hutter R, Klimpfinger M, Sobin L: TNM Atlas, ed 5. Berlin, Springer, 2005.
117 Snijders AM, Nowee ME, Fridlyand J, Piek JM, Dorsman JC, Jain AN, Pinkel D, van Diest PJ, Verheijen RH, Albertson DG: Genome-wide-array-based comparative genomic hybridization reveals genetic homogeneity and frequent copy number increases encompassing CCNE1 in fallopian tube carcinoma. Oncogene 2003;22:4281–4286.
118 Wreesmann VB, Ghossein RA, Hezel M, Banerjee D, Shaha AR, Tuttle RM, Shah JP, Rao PH, Singh B: Follicular variant of papillary thyroid carcinoma: genome-wide appraisal of a controversial entity. Genes Chromosomes Cancer 2004;40:355–364.
119 McIver B, Grebe SK, Eberhardt NL: The PAX8/PPAR gamma fusion oncogene as a potential therapeutic target in follicular thyroid carcinoma. Curr Drug Targets Immune Endocr Metabol Disord 2004;4:221–234.
120 Foukakis T, Au AY, Wallin G, Geli J, Forsberg L, Clifton-Bligh R, Robinson BG, Lui WO, Zedenius J, Larsson C: The Ras effector NORE1A is suppressed in follicular thyroid carcinomas with a PAX8-PPARgamma fusion. J Clin Endocrinol Metab 2006;91:1143–1149.

121 Ain KB, Egorin MJ, DeSimone PA: Treatment of anaplastic thyroid carcinoma with paclitaxel: phase 2 trial using ninety-six-hour infusion. Collaborative Anaplastic Thyroid Cancer Health Intervention Trials (CATCHIT) Group. Thyroid 2000;10:587–594.
122 Zedenius J, Larsson C, Wallin G, Backdahl M, Aspenblad U, Hoog A, Borresen AL, Auer G: Alterations of p53 and expression of WAF1/p21 in human thyroid tumors. Thyroid 1996;6:1–9.
123 Di Loreto C, Tell G, Pestrin M, Pandolfi M, Damante G, Puglisi F: PTEN and Egr-1 expression in thyroid proliferative lesions. Cancer Lett 2005;224:105–109.
124 Kremenevskaja N, von Wasielewski R, Rao AS, Schofl C, Andersson T, Brabant G: Wnt-5a has tumor suppressor activity in thyroid carcinoma. Oncogene 2005;24:2144–2154.
125 Nikiforov YE: Genetic alterations involved in the transition from well-differentiated to poorly differentiated and anaplastic thyroid carcinomas. Endocr Pathol 2004;15:319–327.
126 Oertel J, Oertel Y: Medullary thyroid cancer; in Wartofsky L (ed): Thyroid Cancer: a Comprehensive Guide to Clinical Management. Totowa, Humana Press, 2000, pp 383–388.
127 Marsh DJ, Theodosopoulos G, Martin-Schulte K, Richardson AL, Philips J, Roher HD, Delbridge L, Robinson BG: Genome-wide copy number imbalances identified in familial and sporadic medullary thyroid carcinoma. J Clin Endocrinol Metab 2003;88:1866–1872.
128 LiVolsi VA, Baloch ZW: Follicular neoplasms of the thyroid: view, biases, and experiences. Adv Anat Pathol 2004;11:279–287.
129 Cornille F, Wecker K, Loffet A, Genet R, Roques B: Efficient solid-phase synthesis of Vpr from HIV-1 using low quantities of uniformly ^{13}C-, ^{15}N-labeled amino acids for NMR structural studies. J Pept Res 1999;54:427–435.
130 Younes N, Robinson B, Delbridge L: The aetiology, investigation and management of surgical disorders of the thyroid gland. Aust NZ J Surg 1996;66:481–490.
131 Papotti M, Rodriguez J, De Pompa R, Bartolazzi A, Rosai J: Galectin-3 and HBME-1 expression in well-differentiated thyroid tumors with follicular architecture of uncertain malignant potential. Mod Pathol 2005;18:541–546.
132 Rigaud C: Papillary carcinoma of the thyroid: development of the histological criteria for diagnosis. Study of 29 cases and review of the literature. Ann Pathol 1988;8:211–219.
133 Herrmann ME, LiVolsi VA, Pasha TL, Roberts SA, Wojcik EM, Baloch ZW: Immunohistochemical expression of galectin-3 in benign and malignant thyroid lesions. Arch Pathol Lab Med 2002;126: 710–713.
134 van Hoeven KH, Kovatich AJ, Miettinen M: Immunocytochemical evaluation of HBME-1, CA 19-9, and CD-15 (Leu-M1) in fine-needle aspirates of thyroid nodules. Diagn Cytopathol 1998;18: 93–97.
135 Antonelli A, Miccoli P, Fallahi P, Grosso M, Nesti C, Spinelli C, Ferrannini E: Role of neck ultrasonography in the follow-up of children operated on for thyroid papillary cancer. Thyroid 2003;13: 479–484.
136 Pacini F, Molinaro E, Castagna MG, Agate L, Elisei R, Ceccarelli C, Lippi F, Taddei D, Grasso L, Pinchera A: Recombinant human thyrotropin-stimulated serum thyroglobulin combined with neck ultrasonography has the highest sensitivity in monitoring differentiated thyroid carcinoma. J Clin Endocrinol Metab 2003;88:3668–3673.
137 Cailleux AF, Baudin E, Travagli JP, Ricard M, Schlumberger M: Is diagnostic iodine-131 scanning useful after total thyroid ablation for differentiated thyroid cancer? J Clin Endocrinol Metab 2000;85:175–178.
138 Mazzaferri EL, Kloos RT: Is diagnostic iodine-131 scanning with recombinant human TSH useful in the follow-up of differentiated thyroid cancer after thyroid ablation? J Clin Endocrinol Metab 2002;87:1490–1498.
139 Haugen BR, Ridgway EC, McLaughlin BA, McDermott MT: Clinical comparison of whole-body radioiodine scan and serum thyroglobulin after stimulation with recombinant human thyrotropin. Thyroid 2002;12:37–43.
140 Robbins RJ, Chon JT, Fleisher M, Larson SM, Tuttle RM: Is the serum thyroglobulin response to recombinant human thyrotropin sufficient, by itself, to monitor for residual thyroid carcinoma? J Clin Endocrinol Metab 2002;87:3242–3247.
141 Wang L, Robbins R, Feldman E, Qualey R, Fleischer M, Tuttle R: Management of low measurable thyroglobulin (Tg) levels in thyroid cancer survivors who have negative whole body scans. 76th Annual Meeting of the American Thyroid Association, Vancouver, 2004.

142 Durante C, Haddy N, Baudin E, Leboulleux S, Hartl D, Travagli JP, Caillou B, Ricard M, Lumbroso JD, De Vathaire F, Schlumberger M: Long-term outcome of 444 patients with distant metastases from papillary and follicular thyroid carcinoma: benefits and limits of radioiodine therapy. J Clin Endocrinol Metab 2006;91:2892–2899.
143 Ceccarelli C, Pacini F, Lippi F, Elisei R, Arganini M, Miccoli P, Pinchera A: Thyroid cancer in children and adolescents. Surgery 1988;104:1143–1148.
144 La Quaglia MP, Black T, Holcomb GW 3rd, Sklar C, Azizkhan RG, Haase GM, Newman KD: Differentiated thyroid cancer: clinical characteristics, treatment, and outcome in patients under 21 years of age who present with distant metastases. A report from the Surgical Discipline Committee of the Children's Cancer Group. J Pediatr Surg 2000;35:955–960.
145 Iorcansky S, Herzovich V, Qualey RR, Tuttle RM: Serum thyrotropin (TSH) levels after recombinant human TSH injections in children and teenagers with papillary thyroid cancer. J Clin Endocrinol Metab 2005;90:6553–6555.
146 Schlumberger M, De Vathaire F, Travagli JP, Vassal G, Lemerle J, Parmentier C, Tubiana M: Differentiated thyroid carcinoma in childhood: long term follow-up of 72 patients. J Clin Endocrinol Metab 1987;65:1088–1094.
147 Ozaki O, Ito K, Kobayashi K, Suzuki A, Manabe Y, Hosoda Y: Familial occurrence of differentiated, nonmedullary thyroid carcinoma. World J Surg 1988;12:565–571.
148 Memon A, Berrington de Gonzalez A, Luqmani Y, Suresh A: Family history of benign thyroid disease and cancer and risk of thyroid cancer. Eur J Cancer 2004;40:754–760.
149 Cooper DS, Doherty GM, Haugen BR, Kloos RT, Lee SL, Mandel SJ, Mazzaferri EL, McIver B, Sherman SI, Tuttle RM: Management guidelines for patients with thyroid nodules and differentiated thyroid cancer. Thyroid 2006;16:109–142.

Prof. Gary L. Francis, MD, PhD
Division of Pediatric Endocrinology, Department of Pediatrics
Medical College of Virginia, Virginia Commonwealth University
PO Box 980140, Richmond, VA 23298-0140 (USA)
Tel. +1 804 828 6703, Fax +1 804 628 0267, E-Mail glfrancis@vcu.edu

Van Vliet G, Polak M (eds): Thyroid Gland Development and Function.
Endocr Dev. Basel, Karger, 2007, vol 10, pp 173–187

Hereditary Medullary Thyroid Carcinoma: How Molecular Genetics Made Multiple Endocrine Neoplasia Type 2 a Paediatric Disease

Gabor Szinnai[a,c], *Sabine Sarnacki*[b], *Michel Polak*[a]

[a]Paediatric Endocrinology and INSERM U845 and [b]Paediatric Surgery, Hôpital Necker-Enfants Malades, Paris, France; [c]Paediatric Endocrinology, University Children's Hospital Basel, Basel, Switzerland

Abstract

Multiple endocrine neoplasia type 2 (MEN 2) is a genetic disorder associated with nearly 100% of lifetime risk of medullary thyroid carcinoma (MTC). MTC is the first tumour of the syndrome to manifest, it shows a nearly 100% penetrance and is the most common cause of death in patients with MEN 2. MEN 2A accounts for over 60–90% of patients with hereditary MTC and is characterized by a combination of MTC, pheochromocytoma and parathyroid adenoma. MEN 2B has a high risk of MTC, pheochromocytoma and includes additional clinical features such as mucosal neuromas, ganglioneuromatosis of the gastrointestinal tract, and a marfanoid habitus. Familial MTC, the third subtype of MEN 2, is characterized by MTC in the objective absence of adrenal and parathyroid gland involvement. The identification of the RET proto-oncogene as the susceptibility gene for MEN 2 has fundamentally changed diagnosis and treatment of the disease since 1993. Availability of genetic screening of at-risk children in MEN 2 kindreds made prophylactic thyroidectomy in asymptomatic mutation carriers possible and genotype-phenotype correlations led to codon-oriented prophylactic surgery. In this context, MEN 2 has become a disease of the young child.

Introduction

In the last 15 years, unprecedented advances in the diagnosis and treatment of hereditary medullary thyroid carcinoma (MTC) have been made: (1) by the identification of the RET proto-oncogene as the susceptibility gene for multiple endocrine neoplasia type 2 (MEN 2) in 1993 [1, 2], (2) by the identification of genotype-phenotype correlations in 1995 and 1996 [3, 4], and (3) by the demonstration of age-related progression of MTC in 2003 [5, 6]. Availability of genetic screening of at-risk children in MEN 2 kindreds made prophylactic thyroidectomy in mutation carriers possible and genotype-phenotype correlations led to codon-oriented prophylactic surgery.

Classification of MEN 2

MTC, a rare calcitonin-producing tumour, was first described in 1959. This tumour was found to be a component of the MEN 2. MEN 2A is defined as the association of MTC, pheochromocytoma and parathyroid hyperplasia/adenoma, while MEN 2B represents the combination of MTC, pheochromocytoma, mucosal and intestinal ganglioneuromatosis, decreased upper/lower body ratio, and marfanoid habitus. The third subgroup of MEN 2, the familial medullary thyroid carcinoma (FMTC), is diagnosed in families with 4 or more cases of MTC in the absence of pheochromocytoma or parathyroid hyperplasia/adenoma. Families with 2 or 3 cases of MTC and incompletely documented screening for pheochromocytoma and parathyroid disease may represent MEN 2A [7]. It has been suggested that these families should be considered 'unclassified' until a definitive distinction between MEN 2A and FMTC can be made [8]. MEN 2A may be associated with cutaneous lichen amyloidosis, and MEN 2A and FMTC with Hirschsprung's disease in rare cases. Clinical findings in the 3 MEN 2 subtypes are summarized in table 1.

All MEN 2 subtypes are inherited in an autosomal dominant manner. All 3 subtypes have a nearly 100% lifetime risk of developing MTC, 70% of which become clinically apparent by the age of 70 years [9]. Penetrance of pheochromocytoma and parathyroid disease are lower (table 1). MEN 2A is the most common subtype and accounts for over 60–90% of MEN 2 cases. Up to 5% of MEN 2 cases are of the MEN 2B subtype, while FMTC accounts for 5–35% of all MEN 2 cases [10]. The prevalence of MEN 2 has been estimated to be 1:30,000, the incidence of MTC at 20–25 new cases/year among a population of 55 million [8].

Clinical Features of MEN 2

MTC in MEN 2A

MTC originates from the parafollicular C cells of the thyroid gland. C cells derive embryologically from the neural crest. MTC represents only 3–4% of all thyroid cancers. MTC is sporadic in 70–75% of cases or occurs as the first

Table 1. Classification of hereditary MTC and penetrance of clinical features of MEN 2 by subtype

Subtype	MTC %	Pheochromocytoma %	Parathyroid disease, %	Number of affected family members
MEN 2A[a]	100	50	20–30	any
MEN 2B[b]	100	50	0	any
FMTC	100	0	0	≥4
Unclassified	100	?	?	≤3

[a]Diagnosis of pheochromocytoma and/or parathyroid disease is required.

[b]Characteristic mucosal neuromas on the lips, tongue, and gastrointestinal tract are required.

manifestation of the MEN 2 syndrome in 25–30% of cases. Sporadic MTC is usually unifocal with a peak incidence in the 5th and 6th decades of life. Hereditary MTC tumours are usually bilateral and multifocal occurring already during the first decades of life [11].

Malignant transformation of C cells in MEN 2 is characterized first by a premalignant diffuse C cell hyperplasia (CCH), followed by the appearance of uni- or multifocal MTC, with or without metastases [12]. CCH is diagnosed histologically by the presence of an increased number of diffusely scattered or clustered C cells. MTC is diagnosed when nests of C cells appear to extend beyond the basement membrane and to infiltrate and destroy thyroid follicles [11]. Metastatic spread of MTC to regional lymph nodes or distant sites such as liver is common in patients who present with palpable thyroid mass and diarrhoea.

MTC represents an endocrinologically active calcitonin-producing tumour. Elevated serum calcitonin is a well-established sensitive and specific marker for CCH and MTC [13]. Before the advent of genetic testing, it was the standard screening tool for members of MEN 2 kindreds [14]. Today, serum calcitonin continues to be useful in the surveillance of patients after total thyroidectomy, where elevated levels indicate residual or recurrent disease [13, 15]. Provocative testing can be performed by use of intravenous calcium or pentagastrin. Plasma calcitonin concentrations are measured before (unstimulated basal level) and at 2 and 5 min after pharmacological stimulation of the C cells. Normal values for basal and peak stimulated levels (<10 pg/ml) have been established for an immunoradiometric assay of calcitonin [13].

The three MEN 2 subtypes differ in the age of appearance of MTC, a direct sign of tumour aggressiveness [7]. MTC in the setting of MEN 2B shows the most aggressive course. MTC is often clinically manifest in the first years of life, and metastatic disease has been observed at the age of 2 years [16]. The malignant transformation of C cells is slower in MEN 2A. In a review of the literature of 42 MEN 2A patients with total thyroidectomy before the age of 6 years, 5% had normal histology, 38% showed CCH, and 57% had MTC. No metastatic disease was observed [5]. MTC in an FMTC setting represents the least aggressive form of hereditary MTC and it has a corresponding older age at onset compared to MEN 2A and MEN 2B.

Pheochromocytoma and Parathyroid Disease in MEN 2

Pheochromocytoma has a lower penetrance than MTC in MEN 2. Pheochromocytoma may be uni- or bilateral in MEN 2 and is suspected among patients with refractory hypertension or hypertensive crises. Malignant transformation is rare [17]. Pheochromocytoma occurs usually after MTC, however its presence should be excluded before thyroidectomy in any patient with hereditary MTC [9]. In a large review of 260 paediatric MEN 2A patients, only 3 developed pheochromocytoma at an age of 13–20 years [5]. Annual biochemical screening is recommended in thyroidectomized patients. Appropriate biochemical screening consists of screening for elevated excretion of catecholamines and catecholamine metabolites, such as norepinephrine, epinephrine, metanephrine, vanillymandelic acid and metavanillic acid in 24-hour urine collections [7].

MEN 2A-related hyperparathyroidism is generally associated with mild, often asymptomatic hypercalcaemia, although hypercalciuria and renal calculi may occur. Annual biochemical screening is recommended for those patients who had not had parathyroidectomy and autotransplantation, starting at the time of diagnosis [7, 18].

Molecular Genetics of MEN 2

The RET Proto-Oncogene

The rearranged during transfection (RET) proto-oncogene is localized on chromosome 10. It contains 21 exons and encodes a protein of approximately 1,100 amino acids. The protein is a transmembrane receptor kinase (RTK), termed RET. Its extracellular portion contains four cadherin-like repeats, a calcium-binding site, and a cysteine rich domain. The intracellular portion contains a typical tyrosine kinase domain (fig. 1), which causes downstream signalling events leading to stimulation of several intracellular regulatory pathways of cell survival, differentiation, and proliferation, e.g. the phosphatidylinositol-3-kinase (PI3K)/v-akt murine thymoma viral oncogene homolog 1 (AKT1) cascade which regulates survival and cell cycle progression [19].

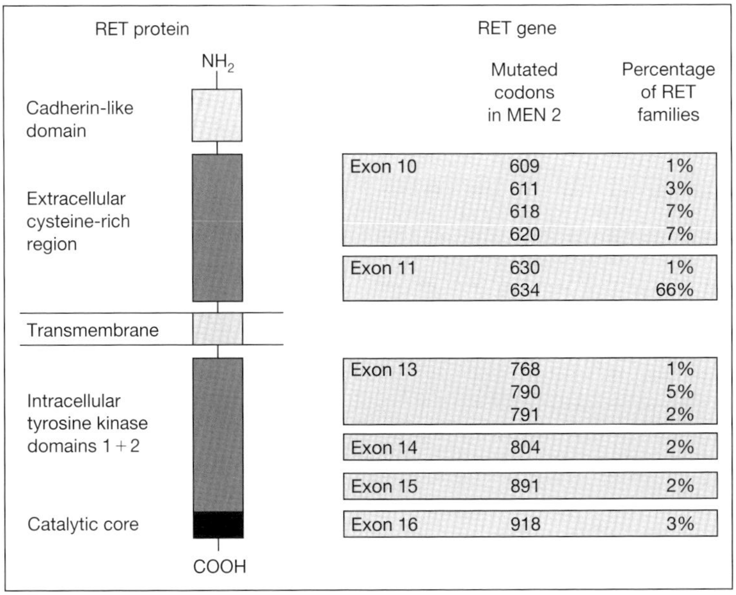

Fig. 1. Schematic representation of the RET tyrosine kinase receptor and localization and frequency of known RET gene mutations in MEN 2. Frequency of the distinct RET mutations in RET families according to EUROMEN 1993–2001.

Under normal conditions, RTK activity is closely regulated. When deregulated, RTKs can become potent oncoproteins. Oncogenic conversion of RTK has been described in the thyroid by two mechanisms. (1) In MEN 2 point mutations lead to a gain-of-function protein responsible for hereditary MCT, pheochromocytoma and parathyroid adenoma. (2) Nonhereditary somatic rearrangements in RET have been identified in papillary carcinoma of the thyroid. Typically, chromosomal inversions or translocations cause recombination of the intracellular kinase-encoding RET domain with heterologous genes, thereby generating RET/PTC chimeric oncoproteins. The result is that follicular thyroid cells aberrantly express the tyrosine kinase region of RET.

MEN 2A syndromes are due to inherited point mutations in the RET gene, leading to a gain-of-function effect, promoting activation of the tyrosine kinase domain. Oncogenic conversion finally results from several intracellular downstream effects, such as enhanced survival signalling and cell-cycle progression, e.g. by the overactive PI3K/AKT1 cascade. Because RET is a proto-oncogene, a single activating mutation of one allele is sufficient to cause neoplastic transformation. The first RET mutations were identified in 1993 in patients with

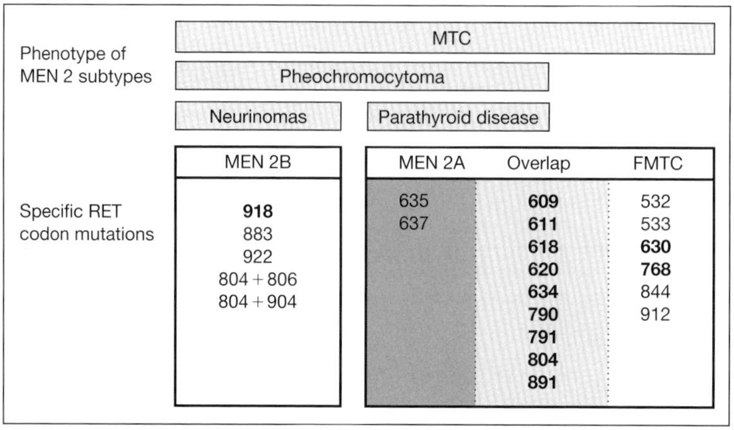

Fig. 2. Phenotype-genotype correlation in MEN 2. All RET mutations described so far correlated to MEN 2 phenotype. The frequent mutations are indicated in bold.

MEN 2A and FMTC [1, 2]. The *RET* germline mutations of MEN 2 are localized in only a small fraction of the open reading frame. Most are in the cysteine-rich portion of the extracellular portion, while some are localized in the tyrosine kinase domain of the intracellular portion.

Oncogenic mutations of RET show a striking and important correlation with the MEN 2 subtype (fig. 2). Approximately 95% of MEN 2A families have a RET mutation in exon 10 and 11. Mutations of codon 634 occur in about 85% of families, mutation of cysteine codons at amino acid positions 609, 611, 618, 620 and 630 together accounts for the remainder of identifiable mutations. Other rare mutations have been reported in single families. Approximately 85% of FMTC families have an identifiable RET mutation. FMTC mutations are evenly distributed among the various cysteine residues (609, 611, 618, 620 and 634) as those of MEN 2A. Why some families with the same mutation will develop only FMTC and not MEN 2A remains to be clarified. Other mutations seemingly specific for FMTC are found in codon 533 of the extracellular cysteine-rich domain and codons 791–891 in the intracellular tyrosine kinase domain of RET. Most MEN 2B patients (95%) carry the M918T mutation in exon 16 in the RET kinase domain. A second mutation A883F has been identified in a small number of families [20].

The mechanisms leading to RET oncogenic conversion in MEN 2 depend on the site of mutation: mutations in codons in the cysteine-rich extracellular domain lead to ligand-independent RET receptor homodimerization and constitutive activation of the tyrosine kinase domain. Different transforming activity between mutations in the cysteine residues has been demonstrated, while different point mutations in the same codon seem not to differ in transforming activity [21]. In contrast, the intracellular mutation M918T responsible for MEN 2B lies within the catalytic core of the tyrosine kinase. It produces constitutive activation of the RET tyrosine kinase by change of substrate specificity, independent of RET dimerization [22]. In line with this model, MEN 2B mutants differ from MEN 2A mutants in the stoichiometry of phosphorylation of tyrosine residues and various intracellular proteins. These molecular differences explain in part the more aggressive tumour biology of MEN 2B MTC.

Genetic Testing and Genetic Counselling

The detection of germline mutations in the RET proto-oncogene had important diagnostic impacts for the management of MEN 2 families: genetic screening of patients at risk allowed the identification of RET mutation carriers with very high specificity and sensitivity and, more importantly, even before onset of disease. Based on these results, and the fact that MEN 2 is a well-described endocrine tumour syndrome, the disease meets the criteria related to indications for genetic testing for cancer susceptibility, outlined by the American Society of Clinical Oncology (ASCO) in its genetic testing policy statement [23]. Current recommendations state that RET mutational screening is mandatory for all children at 50% risk. In consequence, genetic testing has completely replaced biochemical measurements as first-line screening in MEN 2 families [7, 24].

Genetic counselling is essential for patients and their families who face the risk of hereditary MTC. The autosomal dominant transmission mode implies a 50% risk for children of each family member. Ninety-five percent of MEN 2A and FMTC patients have affected parents. Thus, it is appropriate to evaluate the first-degree relatives of an individual MEN 2A patient for manifestations of the disorder. In only 5%, does the disease originate from a de novo mutation or is due to incomplete penetrance of the mutant allele. In contrast, as many as about 50% of MEN 2B patients have de novo RET mutations [25, 26].

The most accurate technique to detect RET mutations is DNA sequencing. DNA sequencing can be done in selected exons with the most common mutations (usually exons 10, 11, 13, 14, 15 and 16) or by sequencing all exons. This is the only technique that allows also identification of new unreported mutations in the RET gene.

Genotype-Phenotype Correlations

Specific RET mutations correlate not only with the expression of specific MEN 2 subtypes, but also with the aggressiveness of MTC within the MEN 2 subtypes [5, 7, 27]. Our meta-analysis of children with MEN 2A provided

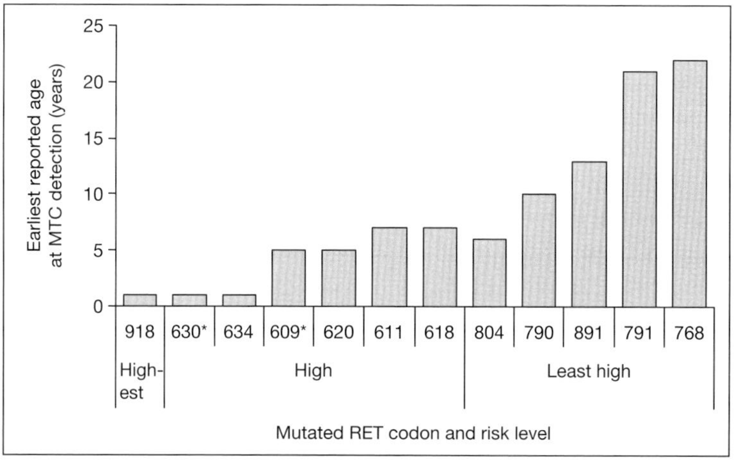

Fig. 3. Earliest reported age at onset of MTC according to mutations in *RET* codons and consensus guidelines risk level. * = Risk levels for mutations in codon 609 and 630 have been changed from least high to high. These risk levels are not covered by the 2001 guidelines but are proposed on the basis of new data [33].

evidence to suggest that subjects carrying c634 mutations were much more likely to present with invasive or metastatic MTC, and more likely to develop persistent or recurrent disease, than were those with c618 or c620 mutations. Yip et al. [27] showed in 71 patients that the three risk groups of *RET* mutations as defined by the 2001 consensus guidelines predicted the MTC aggressiveness in patients with MEN 2. Further, the age of the youngest patient with invasive MCT differs between the different risk levels and the different codons (fig. 3).

On the basis of the aggressiveness of MTC observed in patients with different mutations, RET mutations have been stratified into three risk groups (levels 1–3) in the current consensus guidelines for diagnosis and treatment of MEN type 1 and type 2 (table 2) [7]. Level 3 mutations bear the highest risk for early onset of malignant C cell disease within the 1st year of life. These are 918, 922 and 883 mutations leading to MEN 2B. The youngest patient with metastatic disease was 2 years old [16]. Level 2 mutations (611, 618, 620 and 634) show a slower progression of C cell disease than in MEN 2B, however have a high risk of MTC before 6 years of life. Earliest metastatic disease has been reported in a patient with the 634 mutation at 6 years of age [28], while metastatic disease was found in 13 of 50 patients with the same mutation between 11 and 20 years of age [5]. Level 1 mutations (codon 609, 768, 790, 791, 804 and 891) show the slowest progression to malignant MTC. However, recent studies present conflicting data on the clinical course of MTC in patients with codon 804 mutations.

Table 2. Consensus guidelines for treatment of MTC in MEN 2

MTC risk	Level	RET genotype mutation in codon	Phenotype	Age at total thyroidectomy	Central node dissection
Highest	3	918, 922, 883	MEN 2B	<6 months	yes
High	2	611, 618, 620, 634	MEN 2A, FMTC	<5 years	no consensus
Least high	1	609, 630, 768, 790, 791, 804, 891	MEN 2A, FMTC	no consensus 5–10 years	no consensus

Frohnauer et al. [29] reported one child with metastatic disease at the age of 6 years and in the same series an adult with normal histology in another family. Gimm et al. [30] reported the most variable phenotype of MTC in 23 patients with codon 804 mutations in a series of 140 patients with RET mutations. Lesueur et al. [31] report a low penetrance of MTC in three homozygous and 6 heterozygous patients with V804L and V804M mutations. Both groups propose that individuals heterozygous for weakly transforming mutations of RET require a second germline or somatic mutation in RET or a gene of the RET pathway to result in clinical expression of the disease and that this accounts for the wide clinical variability associated with codon 804 mutations.

The knowledge on genotype-phenotype correlations for all MEN 2 mutations is constantly increasing. They are the basis of current treatment guidelines and for future adaptations.

Management of MEN 2

Total Thyroidectomy

Hereditary MTC is multicentric in 90% of patients, and nodal metastases are present in more than 70% of patients with palpable disease. There is no known effective systemic therapy for MTC. C cells do not concentrate radioactive iodine, and MTC does not respond well to external radiation or conventional cytotoxic chemotherapy. Total thyroidectomy is the only preventive or curative therapeutic approach for MTC [32].

There are two approaches for preventive surgery for hereditary MTC: total thyroidectomy with removal of the posterior capsule alone is the less aggressive intervention. However, the risk of recurrence is higher because the lymph nodes that may be the site of recurrence are not removed. Total thyroidectomy with removal of the posterior capsule and central node dissection is the more complete preventive surgical approach. It removes potential micro-metastasis in the central lymph node compartment, and allows much easier surgical approach of the parathyroids in the case of later development of hyperparathyroidism. In case of advanced disease with palpable tumour further lymph node compartments (jugular) should be resected [32, 33]. Identification of recurrent nerves and parathyroid glands is challenging in very young children and total thyroidectomy requires thus experienced surgeons and an appropriate paediatric environment.

Prophylactic and Genotype-Oriented Thyroidectomy

The advances in the understanding of the molecular basis of MEN 2 syndrome had major impact on management of MTC. First, the availability of genetic screening of at-risk children in MEN 2 kindred made 'prophylactic' or early total thyroidectomy at a preclinical and premalignant stage of disease possible. Second, knowledge on genotype-phenotype correlations for specific RET mutations led to a codon-dependent timing of prophylactic thyroidectomy in mutation carriers.

As outlined above, the current recommendations define three levels of risk for MTC. Level 3 comprises children with MEN 2B phenotype and/or the RET codon 883, 918 or 922 mutations with the highest risk for early development of MTC. They should undergo total thyroidectomy with central node dissection within the first 6 months of life, preferably during the 1st month of life.

Codon 634, 620, 618 and 611 are classified as level 2, with high risk for early and aggressive MTC. Total thyroidectomy with removal of the posterior capsule is recommended for patients with these mutations before the age of 5 years. There was no consensus in 2001 recommendations regarding the need for prophylactic dissection of the central lymph nodes. This issue is still controversial in 2006.

Codon 609, 768, 790, 791, 804 and 891 mutations are classified as level 1, with the least high risk for MTC. There was no consensus in the 2001 recommendations on the management of patients with these mutations with some experts favouring thyroidectomy at 5 years, others at 10 years, and still others on the basis of calcium and pentagastrin-stimulated calcitonin levels.

Based on new data, some authors recommend a change of risk level for codon 609 mutations from the level 1 to the level 2 group and the classification of the codon 630 in the level 2 group (fig. 3) [33]. However, combined genetic and biochemical screening starting in the 1st year of life is mandatory for the detection of unexpected early onset of MTC in all patients independent of the genotype (fig. 4) [33].

Evidence for better long-term outcome after thyroidectomy in asymptomatic RET mutation carriers at a young age is increasing. A meta-analysis comparing the stage of disease and outcome after early (0–5 years) versus late

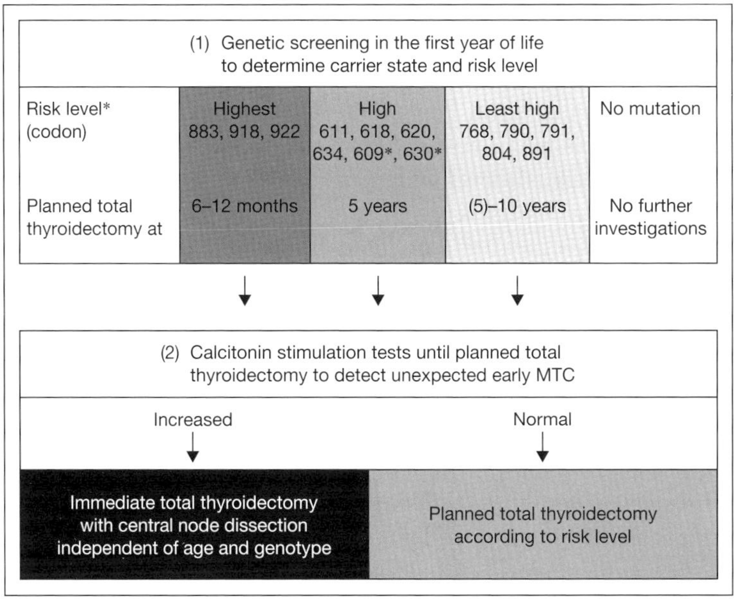

Fig. 4. Combined genetic and biochemical screening to determine timing of thyroidectomy. * = Risk levels for mutations in codon 609 and 630 have been changed from least high to high. These risk levels are not covered by the 2001 guidelines but are proposed on the basis of new data [33].

(6–20 years) thyroidectomy showed significantly lower rates of invasive and metastatic MTC as well as persistent or recurrent disease in patients operated before 6 years of age [5]. Skinner et al. [15] recently presented data from their center on long-term outcome in 50 asymptomatic children thyroidectomized on the basis of proven RET mutations. Basal and stimulated calcitonin levels were normal in 44 patients 5 years or more after thyroidectomy. Four patients had elevation of stimulated calcitonin levels compared to the first postoperative values, however still within the normal range. Two patients showed elevated basal and peak calcitonin levels. All 6 patients with persistent of recurrent disease were 8 years old or older, 2 having positive lymph nodes. However, despite the medical advantage of prophylactic surgery, an extensive discussion and preparation of the families is crucial between genetic diagnosis and planned thyroidectomy.

Management of Patients with Persistent or Recurrent Disease

As outlined above, total thyroidectomy aims to remove the complete C cell mass bearing the gain-of-function mutation. Calcitonin is a well-established sensitive and specific marker for MTC. Calcitonin levels fall usually sharply after thyroidectomy, indicating 'biochemical cure' of an MTC patient. Persistently abnormal calcitonin levels indicate residual disease, however delayed reduction may occur. After postoperative evaluation, all patients should be followed annually by serum calcitonin levels. For this, most authors recommend stimulated calcitonin levels. Elevated basal or stimulated levels during follow-up indicate recurrent disease. Patients with persistent or recurrent disease may stay asymptomatic for many years. A re-operation with formal neck dissection is necessary if nodal metastases are detectable. In the absence of localizable disease by physical or radiological examination without previous neck dissection, re-operation with adequate central and lateral lymph node dissection or observation of the patient by radiological surveillance might be an option [32].

Management of Pheochromocytoma and Parathyroid Disease

Unilateral adrenalectomy appears to be a reasonable management strategy for unilateral pheochromocytoma in patients with MEN 2. The long interval of metachronous pheochromocytoma argues against prophylactic removal of the contralateral normal gland. Patients however need periodic surveillance for the development of the disease in the contralateral adrenal gland. Total adrenalectomy and heterotopic autotransplantation of medulla-free cortex are currently preferred to bilateral adrenalectomy as they may diminish the need for life-long steroid substitution and seem not to expose to recurrence [34]. The coelioscopic approach has rendered this surgical procedure less invasive. Nevertheless, it requires adequate medical preparation, a skilful surgeon and an experienced anaesthesiologist.

Parathyroid disease is not part of MEN 2B. The parathyroid glands may be left in the neck. For MEN 2A and FMTC patients, it is suggested that glands be implanted in the non-dominant forearm to minimize the need for further surgery on the neck after thyroidectomy.

Psychological Impact of Genetic Testing and Prophylactic Thyroidectomy

The medical value of early screening and prophylactic treatment in MEN 2 are contrasted with the psychosocial impact of the genetic diagnosis of MEN 2 on parents and children. Disagreement about the value and the timing of prophylactic surgery between parents of affected children and misconception about genetic disease and the question of guilt are only some of the problems that may cause emotional distress within families. The need for attention to the potential psychological vulnerability of parents and children should not be underestimated.

Perspectives

More Experience with Rare RET Gene Mutations

The current consensus guidelines have been published in 2001 [7]. Meanwhile, more data are becoming available for genotype-phenotype correlations for the rare RET mutations. This knowledge will allow a more precise risk estimation, and more adapted guidelines for prophylactic thyroidectomy (e.g. as proposed for 609 and 630 mutations) and node dissection for the specific risk levels.

New Pharmacological Treatment Option: RET Tyrosine Kinase Inhibitors

Different RET tyrosine kinase inhibitors have been developed in the last years for treatment of malignant haematological disorders and solid tumours caused by tyrosine kinase dysregulation. Preliminary results indicate that different tyrosine kinase inhibitors selectively inhibited cell growth and RET tyrosine kinase activity in MTC cells in vitro in a dose-dependent manner. These results suggest that tyrosine kinase inhibitors might be useful for the treatment of MTC and open a new pharmacological treatment option for MEN 2 patients with persistent or recurrent disease after thyroidectomy [35, 36].

References

1 Mulligan LM, Kwok JB, Healey CS, Elsdon MJ, Eng C, Gardner E, Love DR, Mole SE, Moore JK, Papi L, Ponder MA, Telenius H, Tunnacliffe A, Ponder B: Germ-line mutations of the RET proto-oncogene in multiple endocrine neoplasia type 2A. Nature 1993;363:458–460.

2 Donis-Keller H, Dou S, Chi D, Carlson KM, Toshima K, Lairmore TG, Howe JR, Moley JF, Goodfellow P, Wells SA Jr: Mutations in the RET proto-oncogene are associated with MEN 2A and FMTC. Hum Mol Genet 1993;2:851–856.

3 Mulligan LM, Marsh DJ, Robinson BG, Schuffenecker I, Zedenius J, Lips CJM, Gagel RF, Takai SI, Noll WW, Fink M, Raue F, Lacroix A, Thibodeau SN, Frilling A, Ponder BAJ, Eng C: Genotype-phenotype correlation in multiple endocrine neoplasia type2: report of the International *RET* Mutation Consortium. J Intern Med 1995;238:343–346.

4 Eng C, Clayton D, Schuffenecker I, Lenoir G, Cote G, Gagel RF, van Amstel HK, Lips CJ, Nishisho I, Takai SI, Marsh DJ, Robinson BG, Frank-Raue K, Raue F, Xue F, Noll WW, Romei C, Pacini F, Fink M, Niederle B, Zedenius J, Nordenskjöld M, Komminoth P, Hendy GN, Gharib H, Thibodeau SN, Lacroix A, Frilling A, Ponder BAJ, Mulligan LM: The relationship between specific RET proto-oncogene mutations and disease phenotype in multiple endocrine neoplasia type 2. International *RET* mutation consortium analysis. JAMA 1996;276:1575–1579.

5 Szinnai G, Meier C, Komminoth P, Zumsteg UW: Review of multiple endocrine neoplasia type 2A in children: therapeutic results of early thyroidectomy and prognostic value of codon analysis. Pediatrics 2003;111:E132–E139.

6 Machens A, Niccoli-Sire P, Hoegel J, Frank-Raue K, van Vroonhoven TJ, Roehrer HD, Wahl RA, Lamesch P, Raue F, Conte-Delvolx B, Dralle H; European Multiple Endocrine Neoplasia (EUROMEN) Study Group: Early malignant progression of hereditary medullary thyroid cancer. N Engl J Med 2003;349:1517–1525.

7 Brandi ML, Gagel RF, Angeli A, Bilezikian JP, Beck-Peccoz P, Bordi C, Conte-Delvolx B, Falchetti A, Gheri RG, Libraia A, Lips CJM, Lombardi G, Mannelli M, Pacini F, Ponder BAJ, Raue F, Skogseid B, Tamburano G, Thakker RV, Thompson NW, Tomassetti P, Tonelli F, Wells SA Jr, Marx SJ: Guidelines for diagnosis and therapy of MEN type 1 and 2. J Clin Endocrinol Metab 2001;86:5658–5671.

8 Ponder BA: Multiple endocrine neoplasia type 2; in Vogelstein B, Kinzler KW (eds): The Genetic Basis of Human Cancer. New York, McGraw-Hill, 2002, pp 501–513.

9 Ponder BAJ, Ponder MA, Coffey R, Pembrey ME, Gagel RF, Telenius-Berg M, Semple P, Easton DP: Risk estimation and screening in families of patients with medullary thyroid carcinoma. Lancet 1988;i:397–401.

10 National Cancer Institute: Genetics of medullary thyroid cancer. http://www.cancer.gov/cancertopics/pdq/genetics/medullarythyroid/HealthProfessional/page2.

11 Block MA, Jackson CE, Greenawald KA, Yott JB, Tashjian AH Jr: Clinical characteristics distinguishing hereditary from sporadic medullary thyroid carcinoma. Treatment implications. Arch Surg 1980;115:142–148.

12 Wolfe HJ, Melvin KE, Cervin-Skinner SJ, Saadi AA, Juliar JF, Jackson CE, Tashjian AH Jr: C-cell hyperplasia preceding medullary thyroid carcinoma. N Engl J Med 1973;289:437–441.

13 Niccoli P, Wion-Barbot N, Carron P, Henry JF, de Micco C, Saint Andre JP, Bigorgne JC, Modigliani E, Conte-Delvolx B: Interest of routine measurement of serum calcitonin: study in a large series of thyroidectomized patients. The French Medullary Study Group. J Clin Endocrinol Metab 1997;82:338–341.

14 Gagel RF, Tashjian AH Jr, Cummings T, Papathanasopoulos N, Kaplan MM, DeLellis RA, Wolfe HJ, Reichlin S: The clinical outcome of prospective screening for multiple endocrine neoplasia type 2A: an 18-year experience. N Engl J Med 1988;318:478–484.

15 Skinner MA, Moley JA, Dilley WG, Owzar K, DeBenedetti MK, Wells SA Jr: Prophylactic thyroidectomy in multiple endocrine neoplasia type 2A. N Engl J Med 2005;353:1105–1113.

16 Leboulleux S, Travagli JP, Caillou B, Laplanche A, Bidart JM, Schlumberger M, Baudin E: Medullary thyroid carcinoma as part of a multiple endocrine neoplasia type 2B syndrome. Cancer 2002;94:44–50.

17 Modiglioni E, Vasen HM, Raue K, Dralle H, Frilling A, Gheri RG, Brandi ML, Limbert E, Niederle B, Forgas L, Rosenberg-Bougin M, Calmettes C, et al: Pheochromocytoma in multiple endocrine neoplasia type 2: European study. The EUROMEN Study Group. J Intern Med 1995;238: 363–367.

18 Kraimps JL, Denizot A, Carnaille B, Henry JF, Proye C, Bacourt F, Sarfati E, Dupond JL, Maes B, Travagli JP, Boneu A, Roger P, Houdent C, Barbier J, Modigliani E: Primary hyperparathyroidism in multiple endocrine neoplasia type IIa: retrospective French multicentric study. Groupe d'Etude des Tumeurs à Calcitonine (GETC, French Calcitonin Tumors Study Group), French Association of Endocrine Surgeons. World J Surg 1996;20:808–812.

19 Santoro M, Melillo RM, Carlomagno F, Vecchio G, Fusco A: Minireview: RET: normal and abnormal functions. Endocrinology 2004;145:5448–5451.

20 Marx SJ: Molecular genetics of multiple endocrine neoplasia types 1 and 2. Nat Rev Cancer 2005;5:367–375.

21 Ito S, Iwashita T, Asai N, Muratami H, Iwata Y, Sobue G, Takahashi M: Biological properties of RET with cysteine mutations correlate with multiple endocrine neoplasia type 2A, familial medullary thyroid carcinoma, and Hirschsprung's disease phenotype. Cancer Res 1997;57: 2870–2872.

22 Santoro M, Carlomagno F, Romano A, Bottaro DP, Dathan NA, Grieco M, Vecchio G, Fusco A, Kraus M, Di Fiore PP: Activation of RET as a dominant transforming gene by germline mutations of MEN2A and MEN2B. Science 1995;267:381–383.

23 American Society of Clinical Oncology policy statement update: Genetic testing for cancer susceptibility. J Clin Oncol 2003;21:2397–2406.

24 Lips CJM, Landsvater RM, Höppener JW, Geerdink RA, Blijham G, van Veen JM, van Gils APG, De Wit M, Zewald RA, Berends MJH, Beemer FA, Brouwers-Smalbraak J, Jansen RPM, Ploos van Amstel HK, van Vroonhoven TJMV, Vroom TM: Clinical screening as compared with DNA analysis in families with multiple endocrine neoplasia type 2A. N Engl J Med 1994;331:828–835.

25 Schuffenecker I, Ginet N, Godgar D, Eng C, Chambe B, Boneu A, Houdent C, Pallo D, Schlumberger M, Thivolet C, Lenoir GM: Prevalence and parental origin of de novo *RET* mutations in multiple endocrine neoplasia type 2A and familial medullary thyroid carcinoma. Le Groupe d'Etude des Tumeurs à Calcitonin. Am J Hum Genet 1997;60:233–237.
26 Carlson KM, Bracamontes J, Jackson CE, Clark R, Lacroix A, Wells SA Jr, Goodfellow PJ: Parent-of-origin effects in multiple endocrine neoplasia type 2B. Am J Hum Genet 1994;55: 1076–1082.
27 Yip L, Cote GJ, Shapiro SE, Ayers GD, Herzog CE, Sellin RV, Sherman SI, Gagel RF, Lee JE, Evans DB: Multiple endocrine neoplasia type 2: evaluation of genotype-phenotype relationship. Arch Surg 2003;138:409–416.
28 Gill JR, Reyes-Mugica M, Iyengar S, Kidd KK, Touloukian RJ, Smith C, Keller MS, Genel M: Early presentation of metastatic medullary carcinoma in multiple endocrine neoplasia, type IIA: implications for therapy. J Pediatr 1996;129:459–464.
29 Frohnauer MK, Decker RA: Update on the MEN 2A c804 *RET* mutation: is prophylactic thyroidectomy indicated? Surgery 2000;128:1052–1058.
30 Gimm O, Ukkat J, Niederle BE, Weber T, Tanh PN, Brauckhoff M, Niederle B, Dralle H: Timing and extent of surgery in patients with familial medullary thyroid carcinoma/multiple endocrine neoplasia 2A-related mutations not affecting codon 634. World J Surg 2004;28:1312–1316.
31 Lesueur F, Cebrian A, Cranston A, Leyland J, Faid TM, Clements MR, Robledo M, Whittaker J, Ponder BAJ: Germline homozygous mutations at codon 804 in the RET proto-oncogene in medullary thyroid carcinoma/multiple endocrine neoplasia type 2A patients. J Clin Endocrinol Metab 2005;90:3454–3457.
32 Quayle FJ, Moley JF: Medullary thyroid carcinoma: including MEN 2A and MEN 2B syndromes. J Surg Oncol 2005;89:122–129.
33 Machens A, Ukkat J, Brauckhoff M, Gimm O, Dralle H: Advances in the management of hereditary medullary thyroid cancer. J Intern Med 2005;257:50–59.
34 Inabnet W, Caragliano P, Pertsemlidis D: Pheochromocytoma: inherited associations, bilaterality, and cortex preservation. Surgery 2000;128:1007–1012.
35 Strock CJ, Park JI, Rosen M, Ruggeri B, Denmeade SR, Ball DW, Nelkin BD: Activity of irinotecan and the tyrosine kinase inhibitor CEP-751 in medullary thyroid cancer. J Clin Endocrinol Metab 2006;91:79–84.
36 de Groot JWB, Links TP, Plukker JTM, Lips CJM, Hofstra RMW: *RET* as a diagnostic and therapeutic target in sporadic and hereditary endocrine tumors. Endocr Rev 2006;27:535–560.

Dr. Gabor Szinnai
Paediatric Endocrinology, University Children's Hospital Basel
Römergasse 8
CH–4005 Basel (Switzerland)
Tel. +41 61 685 65 65, Fax +41 61 685 65 66, E-Mail gabor.szinnai@unibas.ch

Author Index

Subject Index